T0091804

ROUTLEDGE LIBRARY EDITIONS:
URBAN PLANNING

Volume 2

METROPOLIS 2000

METROPOLIS 2000
Planning, Poverty and Politics

THOMAS ANGOTTI

LONDON AND NEW YORK

First published in 1993 by Routledge

This edition first published in 2018
by Routledge
2 Park Square, Milton Park, Abingdon, Oxon OX14 4RN

and by Routledge
711 Third Avenue, New York, NY 10017

Routledge is an imprint of the Taylor & Francis Group, an informa business

British Library Cataloguing in Publication Data
A catalogue record for this book is available from the British Library

ISBN: 978-1-138-49611-8 (Set)
ISBN: 978-1-351-02214-9 (Set) (ebk)
ISBN: 978-1-138-47939-5 (Volume 2) (hbk)
ISBN: 978-1-351-06518-4 (Volume 2) (ebk)

METROPOLIS 2000

Planning, poverty and politics

Thomas Angotti

London and New York

First published 1993
by Routledge
11 New Fetter Lane

Simultaneously published in the USA and Canada
by Routledge
29 West 35th Street, New York, NY 10001

Typeset in Garamond by Witwell Ltd, Southport
Printed and bound in Great Britain by
Biddles Ltd, Guildford and King's Lynn

British Library Cataloguing in Publication Data
A catalogue reference for this book is available from the British Library

ISBN 0-415-08135-1
0-415-08136-x

Library of Congress Cataloging in Publication Data

Angotti, Thomas, 1941-
Metropolis 2000 : planning, poverty and politics / by Thomas Angotti.
p. cm. — (Development and underdevelopment)
Includes bibliographical references and index.
ISBN 0-415-08135-1 : $49.95. — ISBN 0-415-08136-X : $14.95
1. Urbanization. 2. Cities and towns–Growth. I. Title.
II. Series.
HT361.A54 1993
307.76—dc20 92-24737
CIP

To my children Antonio, Aida, Jacqueline and Justin

CONTENTS

TABLES AND FIGURES

TABLES

FIGURES

PREFACE AND ACKNOWLEDGEMENTS

While a graduate student of urban planning and tenant activist in the 1960s, I was greatly influenced by the works of C. Wright Mills. Mills spoke out against experts who engaged in 'the scientific applauding of official policies and defaults' (1959: 84). He saw social science as absorbed with vacuous theorizing that obscures reality and 'abstracted empiricism' that breaks it up into innumerable unrecognizable pieces.

When I first began putting together the notes and ideas that evolved into this book, I read a *New York Times* headline about Bhopal, India, where over 2,000 people died and 50,000 were injured due to a poison gas leak from the nearby Union Carbide chemical plant on December 3, 1984.

The headline read:

> Slums Alongside Factories Inevitable, Experts Say

The victims of Bhopal lived in the slums that mushroomed spontaneously over the fifteen years the plant had been operating. Neither the company nor government had planned for the location and development of housing. Union Carbide fought for five years before agreeing to a $470 million settlement with the government, which has recently built housing for a small fraction of the victims.

The 'experts' in the *Times* seemed to characterize everything that is wrong with the prevailing philosophy among urban planners in the West, most of whom are convinced of the futility of planning and the inevitability of social injustice. They have learned to rationalize the failure to plan rationally.

I had entered graduate study in urban planning after working with 'War on Poverty' programs in the United States and the Peace Corps in Peru. These experiences, and the Vietnam War, had brought me to focus my theoretical preparation and research on the questions of social and racial inequality on a global scale and the contradictions of capitalism – to me phenomena that are closely intertwined. After twenty years of work in urban planning as a teacher, activist, consultant and worker in both public and private sectors, this focus seems all the more valid. This book is for me a summation of my experience, an assessment of a world historical process, and a synthesis of the extensive work by countless others who share my same concerns.

Apart from Mills, the greatest influence in my work has been Hans Blumenfeld, an architect, planner and committed activist for social change, a lifelong socialist. It was not just Blumenfeld's pioneering theory of the modern metropolis, which first identified this settlement form as a qualitatively new twentieth-century phenomenon. It was also his breadth of vision and international scope. Blumenfeld viewed the experiences in Europe and the Soviet Union without the ideological prejudice that had made urban planning in the United States so parochial and backward. He also disdained the destruction of neighborhoods by urban renewal and advocated planning on a human scale. His work was relevant to both scholars and to community movements; like all great art it could be understood at different levels by different people.

My greatest disappointment is that Hans Blumenfeld died before I finished the draft manuscript for this book. It would have benefited from his comments and criticisms. I know that until he died in 1988 at age 96 neither his intellectual acuity nor commitment to social change had flagged. Until his last breath, Hans Blumenfeld joined us in demonstrations for peace and remained an advocate for both the smallest urban reforms and the greatest social revolutions. As Peter Marcuse said in his moving review of Blumenfeld's autobiography:

> Blumenfeld's principles of city planning do not fit neatly into any reformist/revolutionary dichotomy. . . . The analysis that underlies Blumenfeld's approach challenges some current simplistic thinking about urban problems on the left. What can practically be accomplished, it suggests, is much more than we conventionally assume; it does not take a revolution to get decent cities. . . . The problems with which he was concerned existed under both capitalism and socialism, and he confronted them under both systems. But the means to their solution might be quite different. Under capitalism, solutions involved clashes of interests, perhaps even class conflicts, rooted in

> fundamentally exploitative and oppressive social patterns. Under socialism as he experienced it, solutions involved overcoming bureaucratic rigidity and conflicting claims on resources at various levels of government, perhaps rooted in equally deep and oppressive patterns. But the solutions themselves were much the same: control automobile traffic, limit the width of buildings, keep different structures in scale, coordinate on a metropolitan level, improve housing for all.
>
> (Marcuse, 1988)

As I labored on early drafts of this book, I began to see a gap in Blumenfeld's work – analysis of the distinct character of the metropolis of developing nations. Blumenfeld identified the universal characteristics of the metropolis, but never looked very closely at the particularities of the 'Third World.' If he had done so, we would have benefited from his great analytic skills in identifying the qualitative distinctions between urbanization in the North and South, and the complexity of the metropolitan development process on a global scale. I have tried to fill this gap, though I feel woefully inadequate in the shadow of Blumenfeld's genius.

A six-month fellowship at the American Academy in Rome gave me the concentrated time needed to finish the manuscript. The quiet and supportive environment at the Academy provided me with an unprecedented feeling of intellectual freedom and the luxury of expansive thinking I could never have with normal work obligations. The library at the Dipartamento di Analisis Economico e Sociale at the University of Venice was opened up for me, thanks to Marco Torres. While in Italy the incredible changes which were occurring in Eastern Europe forced me to reevaluate every previously held proposition about capitalism and socialism and examine more critically the experiences of the Soviet Union, China and other socialist countries. I went through many revisions of the draft. I began to see how the idealist Utopian tradition had been bound up with the planning disasters that were daily revealed in the process of *perestroika*. Indeed, I was further convinced that the professions of architecture and planning everywhere in the world have bought into a formalistic and deterministic approach to planning, though in the West planners usually subjugate their ideals to the logic of profit. In every region, urban planners have exalted an environmentally unsound model of metropolitan growth centered on sprawled development, the automobile and wasteful consumption.

This book is an attempt to present an alternative approach to urban development and planning for the current generation of students in all social science disciplines who, like myself, continue to search for clarification of the main planning issues and what to do about them. It is also for the scholars, teachers and activists I have worked with over the years, whose ideas have helped shape the present work. Indeed, I do not pretend to offer any unique theories (there are none), only to synthesize the collective experience and ideas of those who share the vantage point of people living in neighborhoods in metropolitan areas that are unequal, unplanned and polluted.

While I attempt to provide documentation and evidence to support the propositions I have advanced in this book, it is not an empirical work. My main purpose is to put forth a comprehensive organic analysis and clarify and explain where confusion reigns. I have tried to be true to reality and social experience but this is as much an overall perspective on planning in big cities as it is a documentation of their reality. With Mills, I believe good social science ought to enable us 'to understand the larger historical scene in terms of its meaning for the inner life and the external career of a variety of individuals' (Mills, 1959: 5). It should help us become advocates of the possible and not, like the Bhopal experts mentioned above, the apostles of 'the inevitable.'

I would like to acknowledge in particular the useful comments, patience and commitment to this book by Series Editor Ray Bromley. His criticisms and observations led to significant changes in the final manuscript. I bear full responsibility, of course, for the remaining shortcomings. Reviews of portions or all of the draft manuscript by Josef Gugler, Gavin Kitching, Peter Marcuse, Kosta Mathey and Morris Zeitlin are appreciated. Domenico Cecchini, Sachi Dastidar, Pietro Garrau, Norman Krumholz, Maurizio Marcelloni, Giuseppe Roma and Marco Torres were also helpful in my research. It is with much regret that my friend and colleague Matt Edel died before he could read the manuscript. Many others will also miss his brilliance. Finally, thanks to my wife Emma for her love and support.

In *Metropolis 2000* I have attempted a global overview of metropolitan development and planning in the twentieth century – the century of the metropolis. The point of departure is the classical theory of the modern metropolis by Hans Blumenfeld. In further elaborating this theory, I attempt a sweep and breadth common to leading analysts of the metropolis like Lewis Mumford, Jean Gottmann, Anthony Sutcliffe and Peter Hall.

The fundamental premise is that metropolitan growth is an historically progressive phenomenon that coincides with significant improvements in the material welfare of the world's population. Also, the metropolis is a qualitatively distinct settlement form that requires, but has yet to evoke, qualitatively distinct methods of planning.

Contrary to a bias prevalent in the developed world, the majority of the world's metropolitan areas are in developing nations. Another popular myth is that too much of the world's population lives in metropolitan areas. This myth is often advanced as part of a doomsday theory which depicts the world as overly urbanized, and large and dense cities as pathological problems by definition. The best available data indicate instead that only 20 per cent of the world's population lives in metropolitan areas, and only about 33 per cent live in cities over 100,000. While the metropolis now dominates in economic and political terms around the world and is the predominant settlement form in one continent (North America), the historic contradiction between city and countryside is still a powerful force in the world. In the twentieth century, inequalities between city and countryside have been gradually supplanted by the inequalities among and within metropolitan areas, which have become the 'urban problems' of the modern era.

Metropolis 2000 offers an analysis of metropolitan development and planning in all parts of the world and under different economic and environmental conditions. In the first four chapters, I examine metropolitan development characteristic of the United States, the Soviet Union, and the 'dependent metropolis' of the Third World. In the last three chapters, I examine the problems of urban planning theory and practice in the metropolis and its communities.

The three categories of metropolis reflect general historical tendencies and planning models. They are not rigid ideal types but theoretical categories that synthesize extremely complex historical phenomena. In reality, the urban system in every country has elements of all three, though one is usually dominant.

The U.S. metropolis is an unequal metropolis based on the segregation of land uses and the fragmentation of social groups and political institutions. It is characterized by a densely developed central business district and sprawled suburbs. It is auto dependent, has a highly mobile population, and consumes relatively large amounts of non-renewable energy resources. Planning is determined by the interplay of auto and petroleum monopolies and local real estate interests. It is the classical expression of twentieth-century capitalist urban development.

The Soviet metropolis has been characterized by a relatively integrated social and political structure, limited social mobility and consumer choice. It has an administrative/residential center and relatively high density suburbs. Mass transit tends to predominate. Planning follows a centralized administrative/command structure and tends to reproduce centrally-determined plans. It is the classical expression of twentieth-century socialist urban development.

Somewhat distinct are the metropolitan areas of *Europe*, which have characteristics of both the Soviet metropolis and, increasingly, the U.S. metropolis. European cities tend to be more integrated in both social and land-use terms than the U.S. metropolis. They are smaller, more compact and less sprawled than U.S. cities, and are often part of tightly-knit polycentric regional agglomerations of both large and small cities that include centuries-old historic centers. Urban planning is consistent with the highly regulated capitalist development in which social ownership often plays a significant role. European metropolises, and to some extent Japanese metropolises, are generally expressions of mixed economies and both pre-capitalist and socialist influences.

The dependent metropolis reflects the particular history and dynamics of metropolitan growth and planning in the developing countries of Africa, Asia and Latin America. While the differences within the Third World, and within each region, are substantial, their most acute urban problems stem from dependency on the developed capitalist world and the mass character of poverty. The dependent metropolis mirrors the development of the cities of the former colonial powers as well as the predominant metropolis of the twentieth century – the U.S. metropolis. The planning strategies promoted by the centers of world capital and adopted in the Third World usually reproduce the unequal relations that are at the heart of their metropolitan problems. Planning strategies that have been tried include growth poles, applied central place theory, capital city development, local government autonomy and decentralization, and self help. Case studies of China and Cuba in Chapter 3 illustrate different strategies within planned economies for balancing urban inequalities that have resulted from dependency.

In the last three chapters, *Metropolis 2000* attributes the failure of metropolitan planning to the continued application of ideas and methods developed in the pre-metropolitan era of the industrial city. Metropolitan planning tends to be based on simplistic notions of master planning and master building that ignore the complex division of labor and functions characteristic of the metropolis. It is strongly influenced by Utopian thinking and philosophical idealism.

Metropolitan-level planning must be the starting point for addressing the problems and opportunities of metropolitan areas. However, the missing element in metropolitan planning and governance is neighborhood-level empowerment. Neighborhood planning is yet undeveloped in most metropolitan areas, and where it is most developed, in the U.S. metropolis, it is based on exclusion and segregation and not integration of functions and people at a regional level.

Neighborhood-level planning is the key to integrating the household with the metropolis, residence and workplace. Women have played a pivotal role in pioneering neighborhood political action and community development, and it is no coincidence that the seminal works on neighborhood planning have been by women.

Metropolitan- and neighborhood-level planning need to be integrated through a new set of political and social institutions, a new synthesis of state and civil society. One problem is that the main forms of local governance around the world are leftovers from previous eras. National and municipal governments are becoming less functional as international and metropolitan forces predominate, yet they still wield most power. Metropolitan-level governments, where they do exist, tend to be controlled by national government, and neighborhood-level governance tends to be a mere appendage of municipal government.

In the U.S. metropolis, metropolitan planning is hampered by traditional anti-urban and anti-planning biases. The rejection of planning is a consequence of the hegemony of pro-growth blocs which establish new urban development as a panacea for urban ills.

Too often official planning in the United States is based on anti-urban prejudices which dictate that only a dose of the countryside can cure the city. On the other hand, in planned and mixed economies where metropolitan planning is most developed, it has been formalistic and burdened by a blind physical determinism.

A new progressive approach to planning is evolving. It is based on a vision of metropolitan development as a necessary accompaniment to social development. The metropolis should be seen as a public resource to be preserved for the use and benefit of all people, an inclusionary city. It should be a place where people have maximum mobility without the threat of displacement, and where the greater social efficiency, equity and quality of life made possible in metropolitan areas may be maximized. Neighborhood-level political power and planning is central to progressive metropolitan planning.

At present the image of the privatized 'free-market' U.S. metropolis has emerged as the most attractive and influential on a global scale. It has gained new proponents in the former socialist bloc, in Europe and the Third World. However, the reality is that the mixed economy, and with it a model of mixed public–private planning, remains the most common throughout the world. The mixed economy – not the Reagan–Bush model of deregulated monopoly capitalism nor the Stalin–Brezhnev model of administrative-command socialism – has proven to be more productive, stable and environmentally efficient.

A new democratic and socially responsible image of metropolitan development and planning is emerging out of the synthesis of private and social capital. In the twenty-first century, international monopoly capital may continue to play a determining role in urban growth and planning throughout most of the world, but it will increasingly find itself constrained by the newly emerging global forces: a vastly restructured and decentralized Russia and former Soviet republics, an increasingly diverse and powerful 'Third World' able to demand a new and more equitable international order, and the new progressive social and environmental movements in Europe, Japan and the United States.

In the twentieth century, global metropolitan development was linked to the improvement of the quality of life for a significant minority of the human race. In the next century, human progress will hinge on the metropolitanization of the majority of the world's population, transformation of the old models of metropolitan development and planning, and the elimination of urban inequality.

What is needed is a new way of thinking about the metropolis and urban planning based on the idea that the metropolis is progressive and desirable and not a plague on humanity. We need to study how metro areas are evolving according to the principle of *integrated diversity*, and recognize that planning for *integrated diversity* does make a difference. The new way of thinking should project the current trends of metropolitan development into the twenty-first century, not as fantastical futuristic visions but as sane scientific assessments of the laws that govern metropolitan development, and that permit us to consciously guide that development.

Note: All translations from Spanish and Italian by the author. Minor portions of Chapter 2 were taken from Angotti (1986; 1987). A portion of Chapter 3 was revised from Angotti (1990).

Brooklyn, New York
April 1992

1

THE CENTURY OF THE METROPOLIS

As you move away from the city, you only go from one limbo to another and you can never leave it.

'The Continuous City' in Italo Calvino, *Le Città Invisibili*

Cities are here to stay and we must do the best we can with them.

Lauchlin Currie, *Taming the Megalopolis*

Among the revolutions of the twentieth century, one of the most fascinating and least understood is the metropolitan revolution – the dynamic growth of large cities. Some call them 'megacities' or 'megalopolises.' They are the giant metropolitan areas like New York, Moscow, London, Mexico City, Bombay, Tokyo, São Paulo, and Beijing. They are urbanized regions with millions of people, powerful international settlements whose economic and political importance exceeds that of any city that existed in the centuries since the first human settlements thousands of years ago.

The metropolis in the twentieth century is not just a larger city, but a qualitatively new form of human settlement. It is larger, more complex and plays a more commanding central role – economic, political and cultural – than the industrial city and town that preceded it. It is not just downtown but more like a collection of towns. The social history of the twentieth century cannot be fully understood without taking into account the emergence of the modern metropolis. Indeed, the metropolis is the city of the twentieth century. As Anthony Sutcliffe (1984: 1) notes, 'the rhythms of the giant cities dominate our globe.' According to Jean Gottmann (1989: 168), 'The modern metropolis is the largest and most complex artifact that humankind has ever produced.'

We are in the age of the metropolis. Everything urban is now discussed with reference to the metropolis. When people refer to urban problems today they generally mean metropolitan problems. In the nineteenth century, urban problems were the problems of industrial cities, which usually did not exceed a population of several hundred thousand and today would more often than not be considered small factory towns.

Unlike the industrial city, town and village that preceded it, the metropolis appears to be a relatively durable settlement form. Many industrial cities languished and died when industry moved out. Many towns and villages disappeared because of the mechanization of agriculture and changes in world markets. Many mining towns evaporated when the mines shut down, and many railroad towns were wiped off the map when the stations closed. No metropolis has ever vanished.

For example, in the 1970s there was a movement of capital and jobs from the older industrial northeast of the United States (the Snow Belt) to the Sun Belt in the south and southwest. This prompted dire predictions of a massive decline of the metropolitan areas in the Snow Belt. This decline did not occur, and many Snow Belt metropolises like Boston and New York are now reviving – although the economic restructuring has taken its toll on low-income and minority communities. The devastating decline that was predicted actually occurred in the mining towns and small industrial cities of the Midwest (such as Youngstown, Ohio), not the metropolitan areas. Many neighborhoods declined, some central cities – and parts of central cities – declined, but the metropolis as a whole did not. Indeed, the decline of central cities was usually accompanied by suburban growth within the same metropolitan areas. In another instance, there were predictions that Silicon Valley, in the San Jose (California) metropolitan area, would evaporate after the recent slump in computer electronics. Growth has slowed, but Silicon Valley today can hardly be called a ghost town, and the San Jose metropolitan area is becoming more diversified while still expanding in population.

The metropolis shows no sign of dwindling in size or importance because it does not depend on a single industry or economic activity. It is not just a company town; it is not a manufacturing town. It reflects a new and complex integration of economic activity, including industry, commerce and services (especially the latter) throughout society. It reflects a new and complex division of labor and consumption, and a new concentration and division of political power. It is an expression in geographical space of all these phenomena, and the growth and accumulation of a social surplus unparalleled in world history.

Despite attempts to halt or reverse metropolitan growth, and despite predictions of a return to rural life, the historic trend toward metropolitanization is a steady one. The metropolis is here to stay.

The metropolis is indisputably a global settlement, an international phenomenon, a nodal point in the international division of labor. It serves multiple social, economic and cultural functions across national and international borders. It is the urban expression of a new interconnected and diverse world. It offers valuable locations for production, commerce and management because it gives producers access to a varied and plentiful supply of labor. It offers an unprecedented diversity in consumer products and services, which are part of the basic conditions for the reproduction of labor in today's world. It offers valuable opportunities for face-to-face contact in a complex array of economic and political institutions, particularly those at the commanding heights.

The value of the metropolis goes far beyond its direct economic usefulness to production, distribution and management. The metropolis has substantial social and cultural value, most notably its extensive and diverse opportunities for human interaction. It offers a variety of social and cultural experiences to every sector of society that a smaller settlement could not support – institutions providing specialized experiences in music, art, dance, and sports, and specialized health care and education. It offers a vast menu of employment possibilities and consumer choices. Even when access to these opportunities is unequal, there is an unprecedented range of choices when compared to smaller settlements.

The modern metropolis became an international phenomenon in the twentieth century, in developed and developing nations, capitalist, socialist and mixed economies. Its rise coincides with the industrialization of the globe, the globalization of monopoly capital, the development of socialism and mixed systems, and unprecedented growth and technological advancement. It transcends the end of colonialism and the transition from capitalism to socialism. There are metropolitan areas on all the continents. It has become a truly universal phenomenon.

This is only half the truth, however. Although the metropolis is both dominant and universal, most of the people in the world still live outside its boundaries. It is easy for urban planners from North America to generalize from their own experience and ignore the fact that, although the metropolis is everywhere, the vast majority of the world's population (80 per cent) still lives in smaller settlements. North America is mostly metropolitanized but accounts for only 5 per cent of the world population, and only 14 per cent of the population living in metropolitan areas (see Table 1.1 and Figure 1.1).

Most of the world's population lives in the developing nations of Asia, Africa and Latin America, where the non-metropolitan population predominates. Perhaps as much as half of the world's population still resides in settlements oriented (though no longer exclusively) toward agricultural production and trade.

The metropolitan revolution is rapidly transforming the classical contradiction between city and countryside. David Harvey notes:

> In the context of the advanced capitalist countries as well as in the analysis of the capitalist mode of production, the urban–rural distinction has lost its real economic basis, although it lingers, of course, within the realms of ideology with some important results.
>
> (Harvey, 1989: 73)

In most of the world, especially in Africa and large parts of Asia, urban–rural distinctions are far more than ideological remnants of the past. In many countries inequalities between urban and rural areas are still defining characteristics of economic life. Increasingly, however, inequalities between metropolitan areas and smaller settlements, and inequalities within metropolitan areas – especially between central cities and suburbs – characterize the urban question.

In exploring this question further, it is important to distinguish more precisely between metropolitan and non-metropolitan settlements, particularly between the metropolis and the industrial city of the past. Few analyses of urbanization make this distinction in any decisive or qualitative sense. A recent work by Emrys Jones (1990) continues a well-established tradition of identifying the metropolis as any large city that has existed throughout urban history. By merging discussions of pre-twentieth-century settlements with the twentieth-century metropolis, one usually ends up overestimating the extent of global urbanization. This allows for the advancement of a doomsday theory of urban growth and an anti-urban philosophy. Doomsday theory warns against the overurbanization of the world and calls for measures to stop metropolitan growth. The merging of the two categories of city and metropolis can also obscure the extent to which

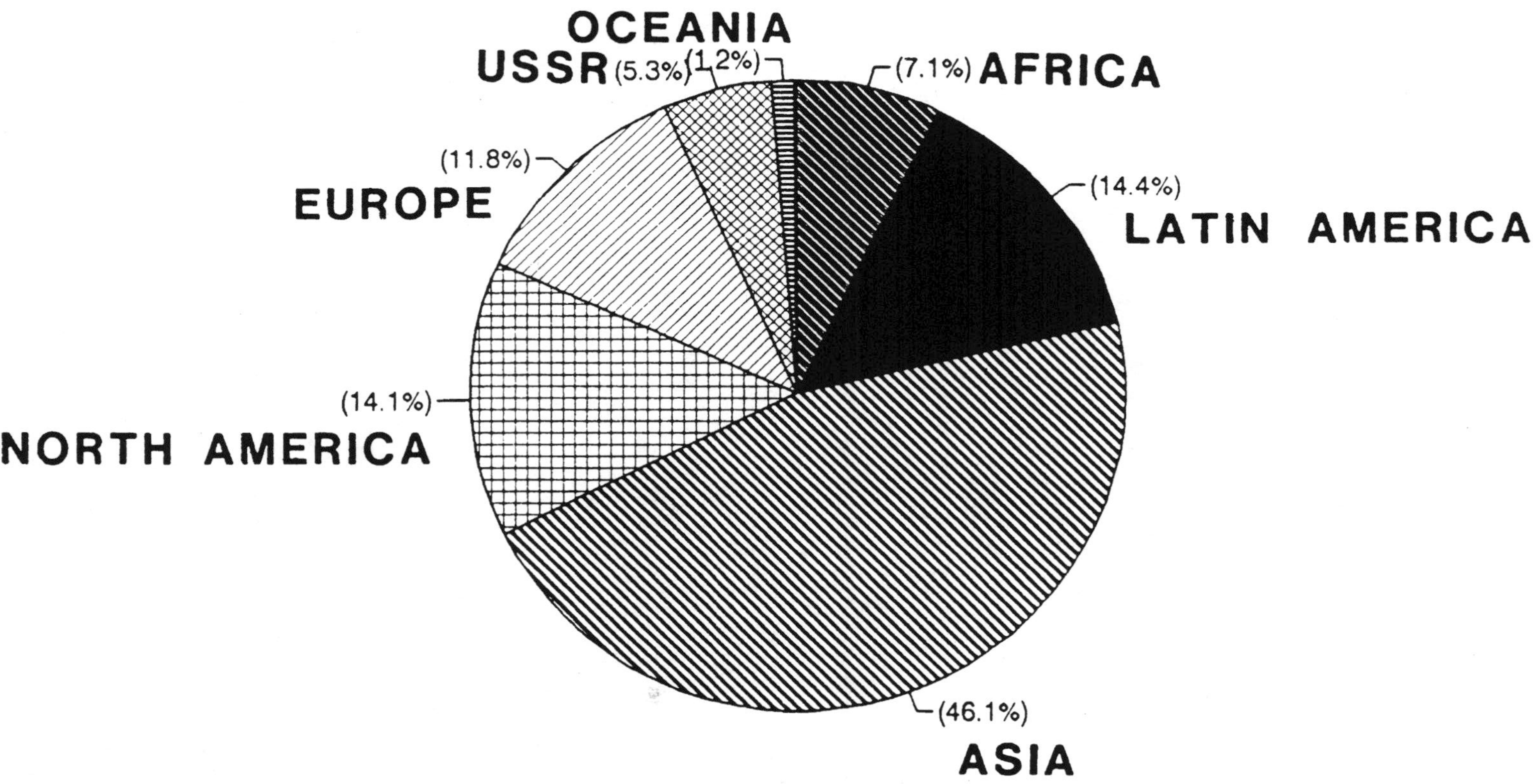

Figure 1.1 Metropolitan population by region.

Table 1.1 Urban population by region

	1990 population in millions			*% of total population*		
	Metro	*City*	*Total*	*Metro*	*City*	*Urban*
Africa	72.3	51.3	645.3	11.2	7.9	19.2
Asia	472.9	235.7	3057.6	15.5	7.7	23.2
Latin America	147.7	66.6	451.1	32.7	14.8	47.5
N. America	144.7	69.4	275.3	52.6	25.2	77.8
Europe	120.6	118.2	498.6	24.2	23.7	47.9
USSR	54.8	83.5	291.8	18.8	28.6	47.4
Oceania	11.8	4.7	26.5	44.5	17.7	62.3
World	1,024.8	629.4	5,246.2	19.5	12.0	31.5

Notes:
Metropolis = >1,000,000 population
City = 100,000–999,999 population
% Urban = % Metropolitan + % City population
Sources: Habitat (1987); United Nations (1989) (adjusted estimates).

the metropolis has come to determine and define the entire urbanization process.

Throughout this book the term *metropolis* (*metropolises* or *metros* in the plural) generally refers to settlements with at least one million population, including central cities and suburbs. A more detailed definition of the metropolis that takes into account other factors besides population size is given later in this chapter. The term *city*, when used in the context of the twentieth century, refers to a medium-sized settlement, between 100,000 and one million population. In this work, cities and metropolitan areas are generally considered to be *urban* settlements, and cities under 100,000 outside the immediate orbit of a metropolitan area are *rural*. This is a broad rule-of-thumb because there are many examples of rural settlements based on agricultural activity with populations above 100,000.

The use of the term *city* in this book differs from popular usage. Frequently people use the term city to refer to the *central city* within a metropolitan area, to distinguish it from the *suburbs*; for example, San Francisco and Chicago are often called cities, even though what is being referred to are actually the central cities within the San Francisco and Chicago metropolitan areas. Frequently people use the term *suburb* to refer to a municipal government or a community with a distinct central business district outside the central city; in this work, however, the suburbs are communities with distinct geographic and social characteristics even if they have no significant commercial center or separate political jurisdiction. Suburbs are basically residential neighborhoods outside the central cities. The term *town* refers here to settlements between 2,000 and 100,000 population. These terms – metropolis, city, town, central city and suburb – refer to geographical phenomena and not to political jurisdictions. Thus, many municipal government boundaries do not coincide with metropolitan areas, cities, central cities or suburbs.

Several urban analysts have focused attention on the metropolis as a qualitatively distinct entity, including Hans Blumenfeld (1971; 1979), Lauchlin Currie (1976), Dogan and Kasarda (1988), Jean Gottmann (1961), Peter Hall (1966; 1989), Lewis Mumford (1961), and Anthony Sutcliffe (1984). These authors have made important contributions to the understanding of metropolitan development and planning. However, their studies either start with an anti-urban bias (Mumford), give little attention to the developing nations of Africa, Asia and Latin America (Blumenfeld; Gottmann and Hall), give little attention to the Soviet experience (all except Blumenfeld), or are mostly collections of individual case studies (Dogan and Kasarda; Sutcliffe). The present work attempts a comprehensive overview of metropolitan development that incorporates and synthesizes many of the contributions of these and many other authors, and seeks to avoid the separate limitations in their scope. The reader will notice, however, the special influence of Blumenfeld's work, and the coincidence of views with Currie.

Doomsday theories of metropolitan growth

It is surprising how few urban theories make a clear distinction between industrial city and metropolis, either in reviewing urban history or in analyzing contemporary urban patterns. It is common for scholars to describe the transformation from industrial city to metropolis as only a gradual evolution, and blur the revolutionary aspects of the transformation. Also lost are the qualitative distinctions between current-day metropolises and the many small and medium-sized cities that have yet to undergo the process of metropolitan transformation.

Lewis Mumford did more than perhaps anyone else to dramatize the problems and potential of large cities, but he also distinguished himself from other urban analysts by promoting the doomsday

Figure 1.2 The Cairo metropolitan region. With over 8 million people, it is the largest in Africa.

outlook. Mumford used terms such as megalopolis and necropolis as hyperboles in his tirades against bigness and technology that vilified every large urban settlement from ancient Rome to modern New York. Mumford preached:

> The dominant world metropolises represented huge concentrations of political, financial, and technological power, developing mainly in this very order: in time they were abetted by religious and educational concentrations of the same magnitude. So effective was this monopoly, so firm this mode of control, so rich its rewards, that they obscured for a time the human penalties of urban congestion: conditions that should have been a badge of shame became almost a mark of honor.
>
> (Mumford, 1961: 551)

He warned of the eventual decline and fall of the metropolis:

> Rapid technological advances in pursuit of obsolete or humanly primitive goals – this is the very nature of the final stage of megalopolitan disintegration, as visible in its day-to-day city planning as in its ultimate plans for atomic, bacterial, and

> chemical genocide. Even the excessive birth-rate may be a symptom of this deterioration. . . .
>
> (Mumford, 1961: 553)

Mumford's disdain for big cities was both a social critique and a prejudice. Mumford saw size and technology as fundamental evils, and lost sight of the differences between the large cities of antiquity and the metropolis of today. While Mumford was ever hopeful the human race would turn away from the misuse of technology, especially weapons of mass destruction, his protestations had the cumulative effect of projecting him as a foe of technology. Perhaps Mumford reflected a deeper trend in political thought that considers the central planning problem to be the control of technology, analagous to Frankenstein's problem of controlling his monster (Winner, 1977). In any case, Mumford's life work was extremely broad, erudite and filled with passionate critiques of modern urban development, so that his anti-urban prejudice appears, on the whole, as an excess and not a deadly flaw (see Hughes and Hughes, 1990).

In a similar vein, Michael Lipton considered that underlying poverty and underdevelopment in the Third World was a decidedly 'urban bias' favoring growth in metropolitan areas. Lipton asserted that the rural poor pay for urban development, and that 'growth and development, in most poor countries, have done so little to raise the living standards of the poorest people' (Lipton, 1977: 14). Lipton believed the key to economic development and the building of a modern industrial society to be 'a mass agriculture of small farmers' (Lipton, 1977: 23).

Lipton's theory follows a common prejudice among development experts who advocate limited urban aid programs based on minimal planning and self-help rather than substantial government intervention. This prejudice against investment in cities is founded on the false premise that improving the urban environment necessarily stimulates urban growth. Lipton's complaint against 'urban bias' is actually the most extreme expression of an anti-urban bias that permeates development literature.

The theories of Mumford and Lipton reinforce many popularly held myths about the metropolis. It is commonly believed that the source of urban problems is the growing number and proportion of people living in cities, especially metropolitan regions. It is said that the earth is becoming one great sprawled metropolis paved from end to end, filled with slums and shantytowns, crime and unemployment, and separated from nature by technology and engineering. The metropolis is considered to be politically unmanageable, environmentally unsound, and humanly unliveable. The solution: stop the growth of the metropolis.

Ever since the beginning of rapid urbanization in the nineteenth century, there has been a profound anti-urban bias in Western thought. From the founding of the United States, national development was equated with agrarian development and cities were perceived as breeders of disease and immorality. In the nineteenth century, Victorians in the United States and Europe preached against the evils of the modern urban age. Anti-urbanism merged with anti-immigrant prejudice as the life style of immigrant workers in the slums of London, Manchester, New York and Chicago was blamed for the terrible living conditions created by industrial capital.

Anti-urban bias is now anti-metropolitan bias, because the large cities of today, the metropolises, are the standard for all urban growth. We are in the age of the metropolis, and anti-metropolitan bias.

The doomsday theory of the metropolis is linked to neo-Malthusian predictions of overpopulation. In discussing 'the population bomb,' Paul Ehrlich warns of 'a serious problem of population distribution,' especially in developing nations where the urban population is growing at a rapid pace (Ehrlich, 1968; see also Velentey, 1977). Doomsday theory is very much a syndrome of analysts in North America, who argue that the way to prevent the kind of urban problems found in the United States from spreading is to prevent other parts of the world from urbanizing. They note that Third World cities and metropolises have most of the problems of the U.S. metropolis and fewer amenities. The metropolis of Africa, Asia and Latin America is seen as a deviant species. The solution is therefore to slow or stop urban growth by means of austere economic policies.

Doomsday theorists propose cutting consumption in the metropolis on the theory that this will slow population growth. In fact, this Malthusian principle has been proven wrong by history. Low levels of subsistence do not curb population growth; if anything, they tend to increase it (Kleinman, 1981). And since the level of income and subsistence tends to be higher in the metropolis (see Currie, 1976: 7), rates of population growth by natural increase tend to be lower. Certainly consumption levels need to be reduced in the small proportion of the world – North America, Europe and Japan – that consumes the largest proportion of the earth's resources and is responsible for the greatest volume of pollution. However, such a measure would probably have little effect on population growth.

The question of urban size is often blown out of proportion in order to promote the doomsday outlook. The very fact that Mexico City will soon reach a population of 25 million is considered automatic proof of the need to stop metropolitan growth. Unproved assumptions about the diseconomies of scale in large agglomerations are asserted as fact. But there is no *a priori* reason why 25 million Mexican people cannot have a better quality of life living in Mexico City than living dispersed among the country's small towns. In fact, they generally do live better in Mexico City, despite all the drawbacks, because the rural areas in Mexico are so much worse off. The real question is whether Mexico can provide a decent standard of living to all, no matter where they live. This requires economic development and planning, not a return to the countryside. But planning is frustrated by Mexico's dependent and unequal role within the international division of labor. Mexican urbanization responds in large measure to the demand for cheap labor and raw materials in the developed nations of the North. Mexico's ability to plan urbanization and improve the urban environment is limited by its debt to Northern banks and reliance on oil exports. Mexico has also become an imitator of the North American model of unequal, inefficient sprawled metropolitan growth.

The latest version of doomsday theory points to the growing environmental degradation around the globe, and calls for limits on metropolitan expansion. This ignores two facts: metropolitan areas are expanding most rapidly in regions that consume the least energy; and the regions that consume the most energy are already more metropolitanized.

As shown in Table 1.2, highly urbanized North America accounts for about 5 per cent of the world population but consumes about 28 per cent of all energy. Africa, Asia and Latin America account for 80 per cent of the world population but consume only 30 per cent of all energy. Per capita energy consumption is highest in North America, the former USSR and Europe, not in the Third World. The most serious problem is not improving energy efficiency in the less metropolitanized Third World – although this is certainly a serious problem – or stopping development in the Third World, but drastically improving energy efficiency in the most developed and metropolitanized regions of the world.

There is no evidence that population growth, whether in urban or rural areas or rich or poor countries, itself causes environmental

Table 1.2 Energy requirements by world region

Region	*Total req.*	*% Total*	*% World pop.*	*Per capita*
Africa	12,142	3.6	12.3	21
Asia	76,802	23.0	58.3	26
S. America	13,601	4.1	8.6	49
N. America	92,847	27.8	5.2	226
Europe	75,062	22.5	9.5	152
USSR	58,991	17.7	5.6	210
Oceania	4,185	1.3	0.5	165
World	333,630	100.0	100.0	66

Notes: Total requirements in thousand terajoules; per capita requirements in gigajoules.
Sources: United Nations (1989; 1989a).

problems (see Kleinman, 1981: 252). The danger of environmental devastation lies not in the development and urbanization of the impoverished nations of the world, but in the further reproduction of the inefficient regime of urbanization that has so far dominated the highly urbanized part of the world, particularly the United States.

The *U.S. metropolis* is an inefficient metropolis, with sprawled suburbs, auto-oriented shopping centers and commercial strips that require high levels of consumption of non-renewable fossil fuels for transportation purposes. The U.S. metropolis uses vast stocks of building materials for its detached single-family dream house. Detached housing requires more energy for heating and home appliances. The sprawled suburb requires that 25–30 per cent of its land be surfaced for streets, sidewalks and parking, increasing the levels of water contamination from surface runoff, which includes lawn chemicals, petroleum products and debris (and in the northern climates, vast quantities of road salts). The U.S. metropolis is ringed and criss-crossed with highways that increase congestion, and despite stringent air quality regulations has air pollution levels that continue to grow because auto use, not mass transit use, is constantly on the rise. The U.S. metropolis extends the working day unnecessarily by requiring lengthy commutation, and puts more people in front of steering wheels and video machines than in public squares or fresh air (see Blumenfeld, 1971: 171–5; Hayden, 1984: 3–62).

The inefficient U.S. metropolis was made possible because at the time the metropolitan revolution began at the turn of the century,

North America possessed seemingly unlimited amounts of land, building materials and energy. This was supplemented by the vast resources available on favorable terms from the dependent nations of Africa, Asia and Latin America. An unequal international division of labor allowed the U.S. to thrive on low-cost energy and labor from across the globe. With the end of colonialism and U.S. hegemony among capitalist powers after World War II, the U.S. was able to capture markets formerly dominated by Europe and Japan. Furthermore, U.S. industry has profited from the export of its auto-centered regime of sprawled development to the developing nations, which suffer the worst consequences of inefficiency because they have the least ability to control or cure the pollution problems that result from it.

The metropolis of Europe, the former Soviet Union and Japan is a much more compact and efficient settlement. Multi-family housing and public mass transit are the rule rather than the exception. The cost of energy to consumers is higher, because fuel taxes cover the social costs of energy consumption and energy prices reflect a more equal exchange relationship with suppliers. It remains to be seen whether this more efficient metropolis will survive as capital becomes increasingly globalized and the auto and petroleum industries are integrated into the various regional regimes of accumulation across the globe. The U.S. model of metropolitan growth has had significant influence in Europe and Japan since World War II. Douglass (1988) shows how the transnationalization of Japanese industry is negatively affecting the quality of urban life in Japan.

Despite the reckless waste of resources in today's metropolis, if we consider the metropolis in all its forms across the globe, it is clear that in general it is the most energy-efficient human settlement form. Rates of domestic consumption of energy for heating, electricity and water are consistently lower in more densely developed areas. Scale economies for infrastructure are greater (although for some services there are clearly diseconomies of scale beyond a certain size). More energy-efficient modes of transportation – rail and trolley, for example – are often not feasible in smaller settlements and achieve maximum economies only in the metropolis.

Contrary to doomsday theory, higher urban densities do not necessarily produce crime and alienation. One need only compare the high crime rates in the sprawled low-density metropolitan regions of the United States with the low crime rates in Europe's and Japan's densely populated metros (see Kleinman, 1981: 248–58). The gross central density of Tokyo and Osaka is about 13,000 persons per square kilometer, compared to about 9,000 in New York City and 4,000 in Los Angeles. Overall metropolitan densities vary considerably from one country to the next. For example, Dallas has only 210 persons per square kilometer, compared to 838 in Paris, 1,875 in Rome and 5,471 in Tokyo (Institut d'Estudis Metropolitans de Barcelona, 1988). Yet no one has yet made a case that crime and other urban problems are dramatically greater in Tokyo and Rome than in New York City, Los Angeles, Dallas and Paris.

Who could argue that the quality of life in Stockholm's high-rise suburbs is any worse than the quality of life in the sprawled freeway suburbs of Los Angeles? Is the social and cultural alienation of the suburban American housewife in a single-family home any less than that of the central city apartment dweller?

Density comparisons are difficult if not impossible in any case. Comparisons often fail to distinguish between gross density (all uses, including land under water) and net density. Net density calculations are not always comparable because they can exclude different uses. Residential densities can include open space on lots or only interior building space. (For discussions of densities see Osborn, 1941; Fischer, Baldassare and Ofshe, 1975.)

A distinction may also be made between daytime and nighttime densities. In metropolitan areas with highly developed Central Business Districts, daytime densities in the center may be anywhere from ten to forty times the nighttime densities (yet more violent crimes tend to be committed at night; this does not take into account white-collar crime, which is less a subject of concern by advocates of urban doom).

Doomsday warnings against high residential density point to overcrowded tenements and high-rise public housing projects as evidence that density causes social problems. Yet the high-rise condominiums of Manhattan's Park Avenue manage to escape scrutiny in these discussions. Certainly no one is proposing to dynamite these fortresses of wealth like the Pruit-Igoe low-income housing project in St Louis, Missouri, which was blown up in 1972 amidst warnings against putting too many people so close to one other.

The doomsday theory of the metropolis often underlies a variety of draconian planning strategies for dealing with urbanization, including attempts to halt or even reverse the process. It has rationalized attempts to bring the city to the countryside by dispersing and decentralizing settlements, and bring the countryside to the city by clearing central urban areas. As we shall see in later chapters, none of these has solved the fundamental problems of urban inequality and environmental quality.

Setting aside the analysis and conclusions of doomsday theory, let us now attempt to explain twentieth-century metropolitan development without an anti-urban bias. In the following section, we examine the evolutionary and revolutionary aspects of metropolitan growth and the qualitative differences in the metropolises of the United States, Europe, Japan, Africa, Asia and Latin America.

TWENTIETH-CENTURY METROPOLITAN DEVELOPMENT: AN OVERVIEW

The evolution of human settlements

The metropolitan revolution occurred after a long and slow evolution of human settlement forms. The earliest human settlements arose in the Neolithic era when sedentary farming yielded a surplus in production, which had previously been limited to producing food for subsistence of the individual clan. As industry and technology developed, greater surpluses were accumulated, and larger settlements evolved as places where the surplus was stored and traded. The first major cities were imperial centers, but the largest of them, ancient Rome, never reached a million population. Until only recently, most of the world's economy depended on agricultural production, and the size and importance of urban settlements were very limited. Even today, most settlements are rural towns and villages.

Generation of a surplus in production has always been the precondition for urban development. Surpluses in agricultural production permitted the growth of commerce and long-distance trading, which in turn stimulated the development of cities as commercial centers. From the fall of Rome to the rise of industry in the sixteenth century, many long-distance trade routes loosely tied together vast settlement networks in Europe, Asia, Northern Africa and the Mediterranean. Industrial production began to evolve five centuries ago with the expanding colonial plunder of raw materials by the European powers, and with the transformation of small-scale urban artisan activities into mass production for the market. Unprecedented reservoirs of capital were accumulated, and in order to survive in the new world the owners of capital had to find ways to reproduce it. By establishing large-scale industrial production in the cities, they encouraged extensive migration from rural areas and the formation of a huge reservoir of labor which, because of its ample supply and relatively low level of material subsistence, was available to work at low wages.

The nineteenth-century industrial and technological revolution marked the climax of this process and brought about a new form of human settlement, *the industrial city*. The industrial city was made possible because of the largest surge in the accumulation of economic surplus ever experienced. Unlike the town and village, the rationale for the growth of the industrial city was large-scale manufacturing for exchange, although the city also incorporated many of the commercial and service functions performed by the pre-industrial towns. Unlike the town and village, the rationale for the growth of the industrial city was the accumulation, reproduction and distribution of capital. The accumulation of capital in the cities enticed peasants off the land as agriculture underwent technological changes that made much rural labor superfluous. The democratic revolutions overthrowing feudalism accelerated and consummated this process.

The term industrial city does not refer to a static ideal settlement type, but corresponds to a particular period in history which came to a close at the turn of the century. It is defined by the dynamic of industrial development that flowered in the nineteenth century, a dynamic qualitatively distinct from pre-industrial development. There are many differences among industrial cities, as there were among pre-industrial cities, and there is no single prototype. This differs from the approach of Gideon Sjoberg (1960), who used the term pre-industrial city as a universal ideal type valid in all historical periods.

The industrial city had large residential sectors for workers, including the surplus labor force, and a small service sector. Even though living conditions were a relative improvement over traditional rural poverty, the cities saw perhaps the greatest concentration of misery since the beginning of time. The pattern of urbanization spread throughout the entire European continent, where the first major industrial cities emerged.

The European pattern of urban development extended to the capital cities of the European colonies, though usually not far beyond. The colonial cities were dependent on the European capitals and became funnels for the export of raw materials, labor and manufactured goods from the colonies, and thus became part of the first large intercontinental network of settlements based on production for exchange. Still, by the turn of the century, the majority of the world population was living in small towns and villages. Industrial cities dominated Europe, and Europe dominated the world, but most of the world outside Europe, including North America, was largely rural.

By the end of the nineteenth century, most of the world's major cities were located along coasts and rivers. This reflected the importance of international trade via shipping. Industrial cities located in the interiors were at a distinct disadvantage, and very few have developed an international stature even today when air and truck freight have taken over a sizeable portion of the shipping industry's business.

Definition of the metropolis

Hans Blumenfeld was among the first to establish the qualitatively distinct character of the metropolis as a new form of human settlement in the twentieth century. Blumenfeld's definition of the metropolis in his classical essay, *The Modern Metropolis*, is an important starting point for our analysis. Blumenfeld notes that:

> from its long, slow evolution the city has emerged into a revolutionary state. It has undergone a qualitative change, so that it is no longer merely a larger version of the traditional city but a new and different form of human settlement.
>
> (Blumenfeld, 1971: 61)

According to Blumenfeld, the industrial revolution in the nineteenth century 'dramatically reversed the distribution of population between village and city.' The major force behind this was the 'dual spur of specialization and cooperation of labor' which 'started a great wave of migration from country to city all over the globe.' This centralizing tendency, however, was soon superseded as the metropolis came into being by:

> an equally powerful centrifugal wave of migration from the city to the suburbs. Although . . . more and more of the population is becoming urban, within urban areas there is increasing decentralization. The interaction of these two trends has produced the new form of settlement we call the metropolis. It is no longer a 'city' as that institution has been understood in the past, but on the other hand it is certainly not 'country' either.
>
> (Blumenfeld, 1971: 64)

According to Blumenfeld, the modern metropolis has a complex of developed districts and open areas, and a well-developed division of functions. The spatial division of functions corresponds with the economic division of labor. There is greater mobility within the metropolis and more opportunity for a wider variety of social and economic activities.

Blumenfeld describes the metropolis as a settlement with a population of at least 500,000. He suggests that one million would be appropriate as a criterion in North America. He states that the population of the metropolis is up to ten times larger than the industrial city, and the land area up to 100 times larger. Obviously, these are not rigid criteria, but Blumenfeld used them to suggest the scale of the metropolis.

In our analysis, the threshold of one million will be used in defining the metropolis. There is a definite logic to this criterion although there is always a certain amount of arbitrariness in establishing a precise figure. Paul and Percival Goodman consider a metropolis to be at least one million population (Goodman and Goodman, 1960: 209). Many United Nations reports focus on 'million cities.' However, there is clearly no universal standard, and in many discussions it is best to establish criteria relative to particular regions or countries.

A settlement much smaller than one million is generally not large enough to have the complex internal division of functions, including a variety of central districts and suburbs, characteristic of the modern metropolis. It does not support a wide variety of specialized economic, social and cultural institutions – for example, a major league baseball team and an opera house. It is not large enough for certain kinds of sophisticated infrastructure. There are very few subways in cities whose population is under one million.

If we use all of Blumenfeld's criteria for defining the metropolis, the determination cannot be simply reduced to a measurement of population level. Therefore, the use of the population criterion in this work is mostly for the sake of parsimony. A thorough analysis of the economic, cultural and spatial characteristics of each individual settlement within its regional context would be required to make a more precise distinction between town, city and metropolis. This is a task beyond the capacity of any single individual or any single study.

We will also follow Blumenfeld's critique of the theory of the megalopolis as a single regional urban unit. Blumenfeld showed how Jean Gottmann's Boston-to-Washington megalopolis is actually a collection of separate metropolitan areas, each growing from the center to its periphery, separated from one another by substantial rural land and open space (Blumenfeld, 1979: 116–28).

Specialization of function

Size is not the only or most significant characteristic that distinguishes the metropolis from other settlement forms. The critical economic and social criterion is specialization of function. Since the metropolis serves a variety of specialized economic and social functions, its livelihood does not depend on a few industries or services. It is the consummate service center, and its services are fundamental to all production and consumption (see Brotchie, Hall and Newton, 1987).

The metropolis exerts a central leadership role in all economic activity. Unlike the industrial city of the past, it does not have to compete with the agrarian sector. In the capitalist world, the metropolis is an international center, the place whence control over capital investment around the world is exercised. In centrally planned and mixed economies, the metropolis also predominates. Everywhere it is the center of decisionmaking and production. Everywhere its dominance is undisputed.

The division of functions within the metropolis corresponds with the divisions within production and consumption. The more complex the divisions between economic sectors, the more complex the division of metropolitan territory. According to Blumenfeld, the metropolis is 'a complex of urban districts and open areas' – no longer just the industrial city, sharply demarcated from its rural hinterland and relatively compact in form. Invariably residential land use predominates in the metropolis, followed by commercial, open space and transportation. In both the market-driven growth of the U.S. metropolis and the centrally planned growth of the Soviet metropolis, there are a variety of residential districts with different housing types, separate shopping districts, government office districts and industrial districts.

In contrast with the older industrial cities, residence and workplace are separate in the modern metropolis. Company housing alongside factories is becoming a scene of the past (though it is still a very live scene in many developing nations – around Bolivia's tin mines, for example). Now, suburban districts tend to be exclusively residential, and even suburban industry is segregated in industrial parks and buffered from residential areas. This marks the end of the evolution from pre-industrial artisan production, in which there was an organic integration of workplace and residence. The epitome of this pre-metropolitan integration was reached in medieval Europe, where artisans had their workshops on the first floor and their residences on the upper floors. This kind of close integration is still present in many Third World cities and metropolitan areas, where the 'informal' service economy prevails (though some form of minimal separation of workplace and residence is still the rule).

One of the chief problems of metropolitan planning today is to reestablish a closer integration between workplace and residence, not necessarily by putting them under the same roof (although that may be desirable in some cases, and to some extent possible), but by planning mass transit, greater flexibility in housing choice and decentralizing economic activities throughout the metropolis. The physical integration of workplace and residence within the same building is neither necessary nor desirable as a rule, though it may be desirable in some circumstances. It may conflict with one of the great achievements of the modern metropolis – mobility. Physical proximity is not the same as spatial integration, nor is it necessary for spatial integration.

The process of urban specialization is fueled by technological innovation. The application of new technology in production and distribution permits and stimulates a more complex division of labor, and spatial function. Particularly important are technological innovations in construction, transportation and communications. The street cars of the late nineteenth century made possible the first streetcar suburbs (Warner, 1962). Underground electric railways made possible more densely developed urban districts. The elevator made possible the Manhattan-style central business district, and the telephone, fax, modem and other electronic advances have made possible the movement to the suburbs of many central business functions. Industrial technology has also proven critical. The introduction of the conveyor belt and assembly-line production made manufacturing in densely developed central cities less economical and promoted the development of large suburban plants in one-story buildings.

The metropolis has the potential for providing greater mobility to individuals than any other form of settlement. This mobility takes three forms: (1) employment mobility; (2) residential mobility; and (3) trip mobility.

(1) *Employment mobility*. The possibility of changing jobs without

changing residence is much greater in the metropolis because there are more jobs of greater variety. Employment mobility is directly related to the mobility of capital, which is in turn a function of the accumulation process. Even when there have been large capital transfers away from the metropolis to stem a decline in the rate of profit, as in the fiscal crisis of the 1970s, new jobs were created as investments were restructured. Mobility may be constrained by a number of factors: strong trade unions, which limit the extent to which capital can move; constitutional guarantees of employment; rigid investment policies; and exclusion due to racial or ethnic discrimination.

(2) *Residential mobility.* Residential mobility often follows employment mobility but not always. Perhaps the greatest residential and employment mobility in the world is to be found in the United States, but there it is accompanied by acute inequalities. In the United States, about half of the population changes residence every five years (Long, 1988: 51). Of all the continents, North America has the highest rate of net inter-regional migration (Habitat, 1987: Table 2).

Residential mobility is not always voluntary. Involuntary displacement is widespread where there are dynamic real estate markets and segregated residential districts, as in the U.S. While there is no accurate count, tenant advocates say that evictions in U.S. cities number in the millions every year. Federally sponsored urban renewal programs displaced millions and gentrification millions more. North America's history is a history of community displacement, of people who have lost their place. Far from the melting-pot ideal, the North American metropolis is an urban place of displaced people who are constantly forced to move. Native Americans were displaced by settlers, African-Americans by slavery, Latin American immigrants by poverty, only to be displaced in their new homes.

(3) *Trip mobility.* In general, there tend to be more opportunities for human interaction and internal travel within the metropolis than in previous urban forms. The need to take more and longer local trips is greater in the metropolis because of the complex division of functions and greater residential and employment mobility. About a fourth of all trips in the metropolis are work-related, followed in importance by shopping and recreational trips. In the developed countries, the full potential for travel within the metropolis can be realized because the transportation infrastructure is usually in place. The maximum travel time between center and periphery may not be forty minutes, as Hans Blumenfeld suggested, but it is often fairly

close to this. In the less developed Third World metropolis, on the other hand, travel times are often double or triple because the transportation infrastructure is inadequate to handle the full volume of traffic, especially at peak flows. It is noteworthy that inadequacy of infrastructure does not necessarily deter people from making trips, nor does it thwart residential or employment mobility. It usually makes trips longer, extends the real working day (and therefore the level of exploitation of labor), aggravates congestion and, because the private automobile and public bus are the main means of transport, causes high levels of pollution.

Accumulation, division of labor and migration

The metropolis, like the settlement forms that preceded it, is made possible by the mobility of capital and labor. Technology advances have facilitated greater mobility, but they alone do not cause it. Migration of labor is one of the two main factors in the growth of the metropolitan population; the other is natural increase (births minus deaths). While migration is the main source of population growth in the early stages of metropolitan development, natural increase is usually more significant in the growth of more mature metropolitan areas.

The main impulse to urban migration has historically been economic. One of the central laws of urban development is the law of capital accumulation. Capital attracts labor, and as capital accumulates so does labor. Capital accumulation produces:

> a relatively redundant population of laborers . . . a surplus population. . . . Capitalist production can by no means content itself with the quantity of disposable labor-power which the natural increase of population yields. It requires for its free play an industrial reserve army independent of these natural limits.
>
> (Marx, 1967: 635)

This general law explains the rapid migration to cities since the onset of the industrial revolution in Europe, the impoverishment of small-scale agriculture, and the creation of a vast reserve army of labor in the metropolis. Immigrant slums, shantytowns and homelessness are therefore an integral part of the process of capitalist development, not an incidental by-product or a separate problem of 'underdevelopment.'

The reserve army of labor drives down wages and the overall living standard of the poorest sectors of the labor force. It allows capital to suppress the wage and living standards of the entire working class. The quality of the urban environment is essentially a part of the overall living standard.

In the twentieth century, the accumulation process changed qualitatively when it spread to every corner of the world. With the rise of giant monopolies and transnational corporations a new international division of labor developed. Industrial and financial capital merged, and investments by transnational corporations around the globe had a decisive impact in transforming the economies of Third World nations (see Cohen, 1981). The global network of transnationals has helped create and reproduce the global network of metropolitan areas (Smith and Feagin, 1987). In Africa, Asia and Latin America, even minimal capital investment by transnational corporations 'pulled' the rural population into metropolitan areas, and control of agroexport industries 'pushed' them even harder.

The law of capital accumulation applies in socialist countries and mixed economies, where the accumulation of capital is also a central factor in urban growth. However, accumulation of *social capital* has generally not been as significant a draw, mostly because there is no reserve army – although employed labor may be stratified and less productive. As a result, the rate and level of urbanization in centrally planned economies has been somewhat less dramatic (see Chapter 4).

The initial high rates of growth in metropolitan areas due to migration tend to stabilize and eventually dissipate as the metropolis gets older. Developed metropolitan areas tend to have fairly low net in-migration. Most of their growth is due to natural increase, and this rate tends to be similar to, or lower than, the rates of population growth of non-metropolitan areas in the same countries.

Thus, growth of the U.S. metropolis is relatively stable due to a low rate of natural increase. Migration from the U.S. countryside has all but ceased. Migration from Europe and Japan has dwindled, and the flow from Third World countries is limited by restrictive legislation.

Despite its *laissez-faire* propensity, the U.S. government has always attempted to regulate the flow of labor to insure a reserve army neither so large as to raise the social costs for its maintenance, nor so small as to exert an upward pressure on wage levels. Immigration policy has fluctuated from relatively liberal to restrictive to keep pace with economic trends, but the general principle that labor should follow capital has been the constant throughout U.S. history.

The U.S. and other developed capitalist countries are now restricting migration to their metropolitan areas, and increasingly rely on the maintenance of labor reserves in Third World metros. These reserves are swelled by the continuing migration from rural areas, which conveniently, for capital, increases the labor pool and drives down wages.

A series of push-and-pull factors operate to draw population from rural to urban areas. Principal among these is the pull of capital. Natural disasters and political factors, such as wars and national growth policies, have also played a role in migration at different historical junctures. For example, there are currently two million refugees from the war in southern Sudan in the urban areas of northern Sudan, hundreds of thousands in Ethiopian cities who fled civil wars in Eritrea and Tigre, and millions in Calcutta, Karachi and other South Asian metros who fled the wars and floods in Bangladesh. But at the international level the accumulation of capital is the main factor in migration. Migration due to natural disasters and wars tends to be transitory and, as we shall see in Chapter 3, most attempts to direct migration within nations by political fiat have also been transitory.

Some analysts claim that a qualitatively new stage was reached in the international division of labor, generally about the 1960s and 1970s (Walton, 1985; Henderson and Castells, 1987). The 'new international division of labor' is supposedly rooted in a more flexible regime of accumulation and consumption unlike the 'Fordist' model adopted by capitalism earlier in the century (Hall and Preston, 1988; Schoenberger, 1988; Scott, 1988; Esser and Hirsch, 1989). Manuel Castells (1989) speaks of the 'informational city,' a dual metropolis with an information-based economy on the one hand and an informal economy on the other. Also, David Harvey (1989) speaks of the post-Keynesian city.

These analysts point to deindustrialization and the restructuring of capital as a qualitative turning point. They refer to the proliferation of small production units at the periphery of metropolitan areas in North America, Europe and some parts of the developing world. They point to the growth of new production enclaves, piecework at home, sub-contracting and sweatshops. They also see the rise of high-tech investment at the metropolitan periphery. For example, one analyst estimates there are over 400 technological parks around the world, the internationalization of the Silicon Valley phenomenon (Perulli, 1989). Another likens the process to Toffler's Third Wave

(Illeris, 1987). Many futurists project the ultimate dispersal of the great metropolitan areas into the rural hinterland.

All of these phenomena are real. However, they do not yet appear to have had more than a limited impact on the overall dynamics of metropolitan development. What is dispersal at the periphery of the metropolis but a new wave of suburban expansion? It is less focused around large industrial plants than the earlier wave of suburban expansion, but that only shows this wave is much smaller and industrial plants are much smaller. There may be new satellite centers, or 'edge cities' (Garreau, 1991) with a mixture of industrial, office and residential uses, but these reflect the maturation and redevelopment of the suburbs, not the emergence of a new settlement pattern. The fact that expansion takes place at the periphery instead of the center probably has more to do with the economics of the land market within the metropolis than the global restructuring of capital. Finally, with regard to high-tech industries, it has been shown that these industries attract a very small proportion of the labor force, usually the upper strata, and they do not follow or attract low-wage labor (Markusen, Hall and Glasmeier, 1986).

The metropolis today: population trends

United Nations statistics that show the majority of the world as urbanized are commonly misused to sound the doomsday alarm against the metropolis. United Nations data rest on definitions of 'urban' established by each individual country. Some countries define any settlement over 2,000 population as urban and others use 100,000 as the criterion. Some countries include every provincial capital no matter what its population. Also, many analyses overemphasize the importance of the metropolis by calculating metropolitan over total *urban* population, instead of total population. Thus, one is given a false impression of a world blanketed by huge metropolitan regions (see Hardoy and Satterthwaite, 1986).

Once again, the reality is that only 20 per cent of the world's population live in settlements over one million population. Even if we allow for substantial undercounting and add in medium-sized cities (between 100,000 and one million population), the proportion is still under one-third. (See Table 1.1.) Some of these medium-sized settlements are within the orbit of metropolitan systems, especially in the developed countries. Some are part of regional agglomerations of small cities, some of which may become part of metropolitan areas in

Table 1.3 Number of metropolitan areas and average size by region

	No. metros	% of total metros	Region's % world pop.	Average size ('000)
Africa	35	9.6	12.3	2,067
Asia	153	41.8	58.3	3,091
Latin America	41	11.2	8.6	3,603
N. America	53	14.5	5.2	2,730
Europe	48	13.1	9.5	2,513
USSR	30	8.2	5.6	1,825
Oceania	6	1.6	0.5	1,969
World	366	100.0	100.0	2,800

Sources: Habitat (1987); United Nations (1989) (adjusted estimates).

the future. But many, especially in the developing nations of Africa and Asia, are merely large market centers in rural areas whose livelihoods are only loosely connected to the metropolitan world.

The proportion of population in metropolitan areas is likely to increase to about one-third by the year 2000. The proportion in all settlements over 100,000 population will be about 40–45 per cent. Thus, the century in which the metropolis has risen to dominate all other settlements will come to a close while the majority of the world still resides in non-metropolitan settlements.

This does not take away from the fact that the growth in human settlements with over one million people in the last century has been nothing short of revolutionary, nor does it take away from their overwhelmingly dominant role across the globe. In 1800, Beijing was the only city of one million population in the world (Lawton, 1989) and by 1875 there were only four. By 1990 there were 366 settlements over one million population.

In 1875, the largest urban settlement was London, with just over four million people. Today, the largest settlement is the Tokyo metropolitan area with over 23 million people.

In 1900, only 5.5 per cent of the world population lived in settlements over 100,000 population. Today the proportion is one-third. (The 1900 figure is by Kingsley Davis, as cited in Dwyer (1979: 9).) In 1875, most of the population throughout the world lived in rural areas and smaller settlements. Today, there is no part of the world without a metropolis, and there are some regions in which most of the population lives in metropolitan areas. (See Tables 1.1, 1.3 and 1.4.)

Table 1.4 Twenty largest cities 1875–1990 (population in '000s)

Rank 1990	*City*	*1875 (est.)*	*1990*	*Times inc. 1875–1990*
1	Tokyo	780	23,372	30.0
2	Mexico City	250	20,250	81.0
3	São Paolo	200[1]	18,770	93.9
4	New York	1,900	17,968	9.5
5	Los Angeles	50[2]	13,075	261.5
6	Calcutta	680	12,540	18.4
7	Shanghai	400	11,960	29.9
8	Bombay	718	11,790	16.4
9	Buenos Aires	216	11,710	54.2
10	Seoul	200[3]	11,660	58.3
11	Rio de Janeiro	274	11,370	41.5
12	Osaka-Kobe	N/A	10,686	-
13	London	4,241	10,400	2.5
14	Moscow	600	9,540	15.9
15	Jakarta	45[4]	9,480	210.7
16	Beijing	1,310	9,290	7.1
17	Rhine-Ruhr	N/A	9,252	-
18	Delhi	207[1]	9,130	44.1
19	Paris	2,250	8,680	3.9
20	Cairo-Giza	355	8,640	24.3
Totals		10,504	204,571	19.5

Notes:
1. 1900 estimate
2. 1890 estimate
3. 1881 estimate
4. 1880 estimate

Sources: Chandler and Fox (1974: 329); Habitat (1987); United Nations (1989).

North America is the most metropolitanized continent in the world, with over half of its population in metropolitan areas, and Africa and Asia the least, with 10–15 per cent each. After North America, Oceania, Latin America and Europe follow. (See Appendix.) The United States is the most metropolitanized country in the world, with 55 per cent of its population in metropolitan areas. Two other large nations have more than a third of their populations living in metropolitan areas – Brazil (39 per cent) and Japan (37 per cent). Smaller metropolitan nations include the United Kingdom (40 per cent), Italy (35 per cent), Germany (35 per cent) and Spain (33

Table 1.5 Countries over 25 million population with largest percentage metropolitan populations (1990 population in '000s)

	Metro pop.	*Total pop.*	*% Metro pop.*
United States	136,878	248,429	55.1
Republic of Korea	22,478	44,828	50.1
Argentina	14,117	32,880	42.9
Colombia	13,048	31,820	41.0
United Kingdom	22,313	56,190	39.7
Brazil	58,680	150,368	39.0
Japan	46,148	123,865	37.3
Italy	20,229	57,563	35.1
Fed. Rep. Germany	20,793	60,332	34.5
Mexico	29,803	89,012	33.5
Spain	13,171	39,748	33.1
Iran	14,882	51,259	29.0
Egypt	11,990	52,536	22.8
France	12,670	55,475	22.8
Turkey	10,764	54,647	19.7
Philippines	11,921	60,973	19.6
Poland	7,281	38,513	18.9
USSR	54,755	291,822	18.8
Pakistan	19,164	112,226	17.1
Vietnam	11,134	66,153	16.8
China	178,820	1,123,815	15.9
Nigeria	15,238	113,343	13.4
Thailand	7,380	55,712	13.2
India	81,039	685,185	11.8
Indonesia	20,889	181,539	11.5

Sources: Habitat (1987); United Nations (1989) (adjusted estimates).

per cent) in Europe; Republic of Korea (50 per cent), Argentina (43 per cent), Colombia (40 per cent) and Mexico (34 per cent) in the Third World. Significantly, several populous nations have relatively small metropolitan populations – the former USSR (19 per cent), Pakistan (17 per cent), China (16 per cent), Nigeria (13 per cent) and India (12 per cent). (See Table 1.5.)

The metropolitan areas in Latin America and Asia are the largest in the world. North America and Europe follow. By contrast, the average size of the metropolis in the former Soviet Union is relatively small. This reflects the particular constraints on the mobility of labor and capital that applied in the USSR, and the role of central planning (see Chapter 4).

The raw United Nations data for Asia and Europe clearly underestimate the size of metropolitan areas. It would not be unreasonable to estimate the proportion of metropolitan population in Asia at substantially more than the 15.5 per cent shown in Table 1.1 – say, 20–25 per cent – and to estimate Europe at greater than its 24.2 per cent – say 30–35 per cent. The data for the other regions appear to be more reliable.

Preliminary 1990 census data for the United States indicate that three-fourths of the population live in census-defined metropolitan areas, but many of these are under one million population. Unlike North America and the former Soviet Union, European figures are not necessarily based on reporting by metropolitan-wide areas, and often conform to municipal boundaries. Hall and Hay (1980) believe that over 80 per cent of Europe (and Japan) is metropolitan but use a very broad definition which includes substantial undeveloped and rural land, and many small settlements beyond the peripheries of urban centers. They make the same mistake that Gottmann does with his theory of the megalopolis, and fail to identify the outer limits of urban land development, and the dynamic of urban growth which spreads from the central cities outward, not uniformly within regions.[1]

Regimes of urbanization and models of planning

From the outset it is important to recognize the distinct characteristics of metropolitan growth and planning in the United States, Europe and Japan, the developing nations of the Third World and the former Soviet Union. The images evoked by the metropolis in each case are different, the problems are different, and the approaches to planning are different. While there are universal laws of metropolitan development that shape all of the three, there are significant particularities that distinguish them. World urbanization cannot be understood without making these distinctions.

The *U.S. metropolis* has become the most influential model for metropolitan development and planning in the twentieth century. North America urbanized rapidly in the twentieth century and is now thoroughly dominated by the metropolis, and the North American metropolis thoroughly dominates discussions of the metropolis.

The U.S. metropolis is a fundamentally unequal metropolis based on the segregation of land uses and the fragmentation of social groups and political institutions. It is characterized by a densely developed and powerful Central Business District and sprawled suburbs. It is auto dependent, has a highly mobile population, and consumes relatively large amounts of non-renewable energy resources.

Development of the U.S. metropolis follows the ideal of the single-family American dream house on an individual lot. Although there is no official national urban policy, national social, fiscal and infrastructure policies combine to promote this dream. Suburban growth and its counterpart, downtown renewal, are premised on the maintenance and reinforcement of inequalities between central city and suburbs, and among neighborhoods, based on differences of race and class.

The amount and kind of planning in the U.S. are determined by the interplay of transnational auto and petroleum monopolies, banks, and local real estate and property interests. Together they make up a constellation of forces that play the dominant role in the politics of the metropolis. Land-use regulations – largely zoning and subdivision rules – tend to reflect the collective interests of developers, particularly the larger ones. Comprehensive land-use planning is mostly limited to smaller cities and towns, where it is usually a captive of local property interests and planners are unable to challenge the private market and public monopolies. The U.S. metropolis is thus a relatively pure expression of minimally regulated twentieth-century capitalist urban development.

Urban planning in the United States is dominated by a widespread pro-growth bias that translates into a suspicion of public planning. Metropolitan development policy tends to be fashioned after the immediate needs of special interests loosely tied together in political–economic *pro-growth blocs*. These blocs reflect the interests of property owners and other sectors of the population that stand to benefit from urban growth. They gain legitimacy through municipal home rule, which tends to frustrate efforts at comprehensive metropolitan planning. Local pro-growth blocs interact with an even more powerful national bloc of monopoly capitalists who benefit from – and to a large extent determine the course of – metropolitan development. Pro-growth blocs often compete with one another for capital and infrastructure projects (or against projects that lower property values). At all levels the pro-growth blocs establish new development as a panacea for urban ills and reduce urban planning practice to a role of simply facilitating that growth.

Somewhat distinct from the U.S. metropolis are the metropolitan areas of *Europe and Japan*, which tend to be more integrated in both social and land-use terms than the U.S. metropolis. Europe and Japan are not as metropolitanized as North America; larger proportions of the population live in small- and medium-sized cities (see Table 1.1).

European and Japanese metropolises tend to be more compact and less sprawled than the U.S. metropolis, and are often part of tightly knit polycentric regional agglomerations of both large and small cities that include centuries-old historic centers. The Japanese metropolis tends to be much larger than both the European and U.S. metropolis.

The European metropolis tends to have a more stable historic center and densely developed suburbs. The historic centers in Europe evolved from antiquity through the long period of feudalism and early capitalism, long before urban real estate became big business. Central city redevelopment has been more limited and gradual than in the United States and development in the periphery has been less sprawled. This goes back to the tradition of the fortified medieval city, where preservation and not consumption of land was the highest value, and where building outside the walls, in the suburbs, was governed by the principle of proximity to the central area and not sprawl.

The metropolitan areas of Western Europe are dominated by CBDs but the role of other functions besides business and finance is greater than in the U.S. Automobiles are less prolific and mass transit more plentiful. In general, the auto–petroleum bloc of monopoly capital is not hegemonic, and the role of the state sector in the economy and urban development is much greater. The European metropolis is the metropolis of mature, regulated capitalism, but also has much in common with the centrally planned metropolis of the former Soviet Union. East European metros are even more a mixture, though in the years since World War II their growth has mostly followed the Soviet model.

The mixture of models and images is apparent in the physical appearance of many European urban settlements. For example, Rome displays an historic center with structures from its pre-capitalist past as well as entirely modern suburbs; widespread condominium ownership as well as publicly-subsidized housing estates; chaotic auto traffic as well as a modern subway and pedestrianized historic center.

Many parts of Europe are characterized by very tightly knit polycentric systems. Blumenfeld (1971: 52) identified these as the English Midlands, the Randstad Holland and Rhine–Ruhr regions, to which may be added the Po Valley in Italy (see Cecchini, 1988) and numerous other conurbations. These are agglomerations of small cities and metropolitan areas with distinct boundaries but organically linked in a single regional system. In Central Europe there are many such systems made up of smaller, non-metropolitan settlements – particularly in Czechoslovakia and the former German Democratic Republic.

Urban planning practice in Europe and Japan is consistent with their mixed economic regimes of highly regulated capitalist development. Social ownership often plays a significant role although private ownership of land and the means of production usually predominates. There are many examples of strong national and metropolitan planning institutions, yet local private property tends to prevail. Master plans are much more than schematic ideas, and are often likely to be implemented, yet they are also circumvented by private landowners. Physical and social infrastructure networks are planned through powerful central welfare states, yet many investment decisions are made in the private sector.

There is no indication that the North American pattern of metropolitan development and planning will be fully replicated anywhere else in the world, despite its overwhelming influence. The experience in Latin America probably reflects to a greater extent than any other the North American experience. Latin America is already highly urbanized, and there are now indications that the rate of metropolitan growth is tapering off there as it has in North America. But even in the Americas there are crucial distinctions between North and South stemming from Latin America's colonial legacy and dependent economic status.

The majority of the world's metropolitan areas – about 60 per cent – are located in Africa, Asia and Latin America. The largest number are in Asia, the most populous region of the world, which accounts for over 40 per cent of all metropolitan areas. (See Table 1.3.) Of the twenty largest metropolitan areas in the world, twelve, about 60 per cent of the total, are in Third World regions (see Table 1.4). Since in general the growth rates of Third World metros are slightly higher than in the industrially developed countries, mostly due to the overall higher rate of natural increase in the population, we would expect the proportion of metropolitan population in Third World settlements to grow modestly in the near future, and to eventually level off as they have in developed countries.

We might also note that over the last century the average increase in population of the twenty largest metropolitan areas (see Table 1.4) was about twenty times the original population. This is twice as much as Blumenfeld's estimate of a tenfold increase. However, at the time Blumenfeld made the estimate in the 1960s, he was quite accurate.

The metropolis is growing most rapidly in places where it has been relatively undeveloped, such as Africa. Metropolitan population is growing three to four times as fast in Africa as the total population, but it is starting from a very low base. In most of the world the metropolis is evolving quite slowly, and there is no clear evidence that the growth rate of the metropolis, over an extended period of time, necessarily exceeds the growth rate of other settlements. In fact, the growth rate in the developed metropolitan areas of North America and Europe has been relatively stable for the last two decades.

The *dependent metropolis* reflects the particular history and dynamics of metropolitan growth and planning in the developing countries of Africa, Asia and Latin America. It is the metropolis of dependent capitalism. While the differences within the Third World, and within each region, are substantial, their most acute urban problems stem from dependency on the developed capitalist world and the mass character of poverty.

The dependent metropolis mirrors the development of the cities of the former colonial powers as well as the predominant metropolis of the twentieth century – the U.S. metropolis. The planning strategies promoted by the centers of world capital and adopted in the Third World usually reproduce the unequal relations that are at the heart of their metropolitan problems. They constantly reproduce unequal urban systems which at the same time tend to diminish the relative significance of urban–rural differences. While many Third World metros are loosely planned in the image of the U.S. metropolis – CBD-centered, auto-dependent and sprawled – many use central planning measures and master plans, and have highly developed state sectors. To the extent that their economies are mixed, so are their metros.

Development of the metropolis of Africa, Asia and Latin America is highly dependent on external factors instead of following the logic of internal economic development. It is governed by the movement of capital and consumption patterns propagated by transnational corporations. Its growth is shaped by the unequal relations underlying international capital, and internal inequalities are usually much more pronounced than in the metropolitan areas of the U.S. and Europe. Using relative standards, the majority of urban dwellers in the dependent metropolis live in poverty, while in the U.S. metropolis the minority live in poverty.

The category dependent metropolis includes widely differing metropolitan areas, as diverse as the nations of the Third World. The differences among nations reflect the timing of their transformation from colonies to formal independence, their particular roles within the international division of labor, and the degree to which they have been able to achieve some level of economic independence (see Chapter 3).

The term Third World is used throughout this work only because it has achieved widespread use as a synonym for the less developed countries of Africa, Asia and Latin America. Use of the term does not imply any validity to the theories of three separate worlds. With the internationalization of capital and the demise of administrative/command socialism, distinctions between what were considered the first two worlds have become unclear, and distinctions within the 'Third World' are now more substantial than ever. A sizeable portion of the Third World is as highly developed as some parts of the other two worlds (see Harris, 1986).

The *Soviet metropolis* was characterized by a relatively integrated social and political structure, limited social mobility and consumer choice. It has an administrative/residential center and relatively high density suburbs. Mass transit tends to predominate. Planning follows a centralized administrative/command structure and tends to reproduce centrally-determined plans. It is the classical expression of twentieth-century centrally-planned socialist urban development.

In Eastern Europe and the former Soviet Union over the past forty-five to seventy years, metropolitan areas have developed within the context of central economic plans and master land-use plans. In the absence of a private real estate market, there was a relatively even spatial pattern of development. In the absence of transnational capital, metropolitan growth followed the pattern of social capital investment via the national budget; the metropolis tended to grow more slowly, while small- and medium-sized cities played a significant role.

Soviet city planners introduced many bold planning concepts and practices that have been emulated around the world: integration of uses, comprehensiveness, integration of physical and economic planning, the guiding role of public infrastructure, etc. They also introduced an approach to city-building based on serial reproduction of standard models that belongs on an industrial assembly line instead of in the metropolis (see Chapter 4).

Characteristics similar to the former Soviet Union may be found in socialist countries that have histories as colonies and whose cities were formed under the regime of dependent capitalism. These include countries like China, Cuba and Vietnam. Metropolitan areas in these countries have developed in accordance with both dependent capitalist and socialist planning principles.

The differences in metropolitan development in different regions of the world are not simply differences of size. There are critical ways in which the underlying processes of social, economic and political development differ from one another. While there appear to be universal laws of metropolitan development that hold true for all, there are important differences that separate the industrially developed and developing nations, capitalist, socialist and mixed economies. Even within these categories, there are marked distinctions to be made: for example, between the United States, Europe and Japan; between the more developed countries of Latin America like Argentina and Venezuela and less developed countries like Peru and Bolivia; between China and the former Soviet Union, India and Singapore, Cuba and Vietnam, and so on. In short, the metropolitan world displays a complex mixture of economic and urban development principles that are played out in a variety of historic and geographical contexts.

The terms *U.S. metropolis*, *dependent metropolis* and *Soviet metropolis* more or less correspond with the three main urbanization regimes in the twentieth century. They are not rigid ideal types or geographical categories. They are present in every national urban system, though every national urban system is dominated mostly by one of them. They are present in every region of the world, though in every region one of them predominates. Just as the national economies of the world are increasingly complex mixtures of economic tendencies, so also are national urban systems.

For example, there are some capitalist countries like Sweden whose urban areas resemble European Soviet settlements more than North American settlements. There are some Third World metros like Seoul that look more like North American metros than other Third World metros. Some Third World metros, like Brasilia for example, bear images of Soviet planning, yet some metros of the former Soviet Union, like Tashkent, bear images of Third World metros. However, there is a general correspondence that makes these categories, in the broadest theoretical and historical sense, valid ones for a general discussion of the metropolis.

2

THE U.S. METROPOLIS

Mobility and inequality

In a community where public services have failed to keep abreast of private consumption things are very different. Here, in an atmosphere of private opulence and public squalor, the private goods have full sway.

John Kenneth Galbraith, *The Affluent Society*

Laissez faire, laissez passer.

François Quesnay

The United States is a truly metropolitan nation. Fifty-five per cent of its population live in settlements over one million population, 78 per cent live in settlements over 100,000 population, and less than 3 per cent work in agriculture. The metropolis is indisputably the center of all economic, social and political life.

The U.S. metropolis is not only a significant historical fact, it is an international model. To many countries and individuals throughout the world, both rich and poor, the wealth and glitter of the U.S. metropolis, and the North American approach to metropolitan planning, are to be emulated if not admired. There are the imposing skylines of Manhattan, Chicago and Miami Beach; the neon opulence of Las Vegas; the planned, orderliness of Disneyworld; and the monumental spaciousness of suburban affluence. The image of the U.S. metropolis is that of unprecedented individual wealth, freedom and mobility. But it also projects a distinct image of economic and racial inequality, poverty, crime and violence.

The U.S. metropolis is the most prominent example of urban development unfettered by any deeply rooted pre-industrial urban structure or nineteenth-century industrial urban structure (Australia is another example). As the U.S. entered the twentieth century, the century of metropolitan growth, little remained of its colonial-era settlement structure, and most of the relatively new industrial cities of the nineteenth century were readily incorporated into and engulfed by metropolitan regions. Thus, metropolitan areas in the U.S. are distinctively twentieth-century products in form and appearance – especially in the West and South.

At the turn of the century, the U.S. population was mostly rural, and all major cities were located in the East. Within a few decades, rapid industrial growth, immigration from Europe and settlement of the vast territories in the West radically transformed the urban landscape. By the end of World War II, about half of the population lived in metropolitan areas as defined by the census. By the end of the century over 80 per cent will live in these areas.

By the time of the post-war ascendency of the U.S. to global superpower status, capital was concentrated in the metropolis it had created; by the 1980s small-scale agriculture had been almost entirely displaced by agri-business. Metropolitan areas flourished throughout the nation, and the sprawling new highway metros of the West and South epitomized the rise of the new metropolitan America. (For detailed histories, see Glaab, 1963; Hauser and Schnore, 1965; Reps, 1965; McKelvey, 1968; Callow, 1969; Teaford, 1986; Zeitlin, 1990.)

Metropolitan areas in the U.S. are in many ways similar in economic and physical structure to those of Europe and Japan. They are global metros whose development is intimately tied to the international division of capital and labor (see Sassen, 1991). With the relative decline of the U.S. economic position since the 1970s, the historic differences among the world metropolises may be further waning. Here, however, we will emphasize some of the important differences that have distinguished the metropolitan areas of the United States from those of Europe and Japan throughout most of the century.

The salient features of the U.S. metropolis include: an image of physical opulence and wide consumer choice, social inequality and spatial segregation, a high level of mobility based on auto use, poverty and residential displacement in central cities, and the preeminence of private space. The Central Business District dominates the central areas, and suburbs are sprawled around single-family homes, local shopping malls and highway commercial strips. The U.S. metropolis reflects a regime of urbanization based on a loosely regulated real estate market, the reification of profit, and conspicuous individualized consumption.

In this chapter we will focus on the internal characteristics of the

U.S. metropolis and the role of urban planning. In the last three chapters we will explore in greater detail the theory and practice of metropolitan planning in the U.S. and elsewhere.

THE U.S. METROPOLIS: AN OVERVIEW

Metropolitan growth in the U.S. is the product of a particular kind of accumulation that is minimally constrained by government planning and regulation. While depending on government support and subsidy for its expansion, the urban development community in the U.S. has enjoyed greater independence from state regulation than its European parents, and a national culture and ideology that lauds and subsidizes free enterprise before interfering with it. This has contributed to a free-wheeling entrepreneurial atmosphere for metropolitan development, less restrained by government regulation, historic tradition and pre-capitalist urban remnants.

Urban planning by the public sector, especially comprehensive planning, has played a minimal role in metropolitan development. Consider the comment by architectural critic Witold Rybczynski that the proposals of American planners and architects 'have usually seemed slightly irrelevant and beside the point, and just about as useful as a copy of the Larousse Gastronomique would be to a short-order cook in a roadside hash house' (Rybczynski, 1991).

Among the most important distinguishing features of the U.S. metropolis are the dramatic economic inequalities between districts – between central cities and suburbs, and within central cities and suburbs – and the sprawled auto-centered form of urban growth. These are related to the prominent role of real estate markets and profit, the unique regime of accumulation, reproduction and consumption characterizing the U.S. economy, and the relatively subordinate role of labor in planning for growth. They are related to the historic role of racism in the U.S. In the following sections we shall examine these features in some detail.

Inequality and exclusion

In appearance, the U.S. metropolis seems to be an open city, a sprawling display of wealth without walls, a place with vast opportunities for residential, employment and transportation mobility not found in other parts of the world. However, this apparent freedom is constrained and rationed by economic laws. Class divisions and intraclass stratification contribute to the production and reproduction of homogenous, exclusive residential enclaves, both in central cities and suburbs. What appear to be open communities are in reality unwalled enclaves that seek protection from the integrating tendencies of metropolitan growth.

The wide variety of residential, commercial and industrial districts in the U.S. metropolis translate into greater freedom of choice. However, the choices are available only to those wealthy enough to reject the least attractive options. Social and ethnic diversity among neighborhoods, especially in central cities, is not the result of a harmonious process of integration but the product of discrimination and exclusion based on race, national origin, gender and life style.

The tendency to diminish the role of racism may be found throughout the urban literature. One popular text from the 1950s stated:

> much of the racial friction in cities today has less to do with skin color than the new arrivals' lack of knowledge of such rules of the game as not throwing garbage out of the window.
>
> (*Fortune*, 1958: 28)

The United States is the only major capitalist nation in which capitalism was born concurrently with a system of slavery. Slavery thrived during half of its history as a nation, and for most of the other half local laws in the South were based on the segregationist 'separate but equal' doctrine. With the industrialization of Southern agriculture and migration of Southern blacks to metropolitan areas in the post-World War II years, and with the rise of the Civil Rights movement, Southern Jim Crow laws were finally removed from the books. It was not until the landmark 1954 *Brown v. Board of Education* Supreme Court ruling, however, that the 'separate but equal' doctrine was struck down. Thus, racism has played a significant role throughout U.S. history (see Clark, 1965; Downs, 1970: 75–114; Osofsky, 1971).

Today, the majority of African-Americans live in segregated central city neighborhoods. Ninety per cent of all blacks in the New York metropolitan region, for example, live in the core municipality, New York City, according to preliminary 1990 Census data. While one-fourth of all New York City residents are black, only 17 per cent of the population in the five inner-ring suburbs are black, and only 7 per cent in the outer suburban counties. Within central city and suburban counties, the black population is further segregated by neighborhood, and most of the black population in suburban counties is concentrated

Figure 2.1 Homelessness and inequality.

in the older urban cores, such as Trenton, Bridgeport and New Rochelle (Hanley, 1991). As shown in various studies, the expansion of the black middle class since the 1960s has resulted in only minimal integration of some suburbs (see Orfield, 1986).

Central city and inner suburban populations are becoming more ethnically mixed, as Latino, Asian and other immigrants take the place of the declining white population. However, the physical proximity of different groups does not necessarily imply social and economic integration. The economic condition and mobility of most blacks has not substantially improved relative to their neighbors from other ethnic groups. In central cities, the proportion of the African-American population living below the poverty line is actually increasing (Herbers, 1987).

There is no need to reiterate here proof of the continuing poverty in African-American communities from Washington, D.C. to Watts in Los Angeles (see Auletta, 1982). Whether we use measures of employment, income, infant mortality, homelessness or years of schooling, the gap between the quality of life in the ghetto and the

Figure 2.2 The Central Business District.

Figure 2.3 Freeways and sprawl.

rest of the metropolis has further widened since the U.S. surrendered in the War on Poverty and cut back social spending in the 1980s.

In the 1980s, *homelessness*, which disproportionately affects blacks, became a widespread feature of metropolitan America (see Hopper and Hamberg, 1986). The rise of homelessness coincided with the sharp decline in government expenditures for new construction of subsidized low- and moderate-income housing and social services during the Reagan years.

Violence is another particular feature of the U.S. metropolis. In a nation where the myth of the Wild West and the image of the gun-toting cowboy remain powerful, there are an estimated 200 million guns, enough for at least two per family. Violence is a thriving industry in the metropolis, and feeds the insurance and public safety industries. Because violent crime is perceived to be a chronic problem, all urban planning and design must take into account potential threats of violence. Thus, planning seeks to maximize the use of private space, which is considered safer, and minimize the use of public space. White-collar crime, on the other hand, which breeds in private spaces, is not widely perceived as a chronic problem.

The U.S. metropolis is not the only exclusive metropolis. Many European metros, like London, Paris and Berlin, also have homogeneous, segregated neighborhoods of immigrants from the West Indies, South Asia, Africa or the Middle East. There are racial and ethnic boundaries within many Third World metropolises such as Johannesburg, Lagos, Rio de Janeiro, and Calcutta, though they are often obscured by the mass character of poverty. But the U.S. metropolis is the world model, the metropolis *par excellence* of unamalgamated immigrants and insular communities, the 'melting pot' that never was (see Glazer and Moynihan, 1970).

The U.S. metropolis is fragmented into communities, many of which have their own municipal governments, that strive to be enclaves of single-family dream homes. They seek to preserve their 'character' by excluding low-income housing and the poor, industry (except for 'clean' high-tech industries), waste disposal facilities, and anything that would tend to lower residential property values. Poor communities are easily tempted to abandon these exclusionary goals in order to attract sorely needed jobs and tax revenues. They often acquiesce in their powerlessness to exclude, but they too are influenced by the ethic that equates quality of life with exclusion.

Integration of functions, and integration of residents across economic and racial groups, is not the standard. Integration is associated with low property values: communities that are exclusively residential are valued highly, and communities that are exclusively made up of single-family homes on large lots are valued more highly than those with a more dense configuration (although price per square foot may be higher in more densely developed central locations).

The counterpart of exclusion is 'ghettoization,' or segregation. The system of exclusion would not function if there were not a place to put people who are excluded. Harlem and the South Bronx are products of exclusion from such areas as Park Avenue, Riverdale and upper Westchester County. The South African townships are products of apartheid's exclusion of Africans. Thus, the key to eliminating segregation is not to integrate Harlem or the townships but to eliminate segregation everywhere in the metropolitan area.

Urban inequalities are not created by urban planners, but they are often protected by them. Planning does not have to be discriminatory, and physically separating land uses and social groups is not necessarily tantamount to segregation. However, planning often endorses the existing state of segregation and unequal development.

In this nation of loose regulation and deregulation, public sector urban planners do not have the power or influence to institute social exclusion, but they often reinforce it. The exclusive upper-class enclaves of Westchester County (New York) are protected by a fragmented system of governance and planning that allocates better housing and services to municipalities with higher property values. Planners in these localities are entrusted with the protection of these values and the social privileges that go along with them. They use zoning and subdivision regulations, growth control and historic preservation measures which, when coupled with local taxation and spending policies, effectively shut the door to low- and moderate-income housing. Planners are called upon to insure only 'appropriate' development, that which promotes the 'highest and best use' of the land. Planners do not shape or control development, they codify and legitimize it with their zoning rules. Indeed, zoning is the main instrument of planning; it is used to promote central city development and codify suburban exclusion (see Reps, 1964; Babcock, 1969).

Middle-income and more privileged working-class enclaves in central cities are also protected by ethnic privilege. Take, for example, Boston's North End, a mostly white neighborhood that survived while low-income, minority and integrated neighborhoods all around it were displaced by downtown expansion, highway construction and urban renewal. Then there are rustic havens like Seagate in Brooklyn

(New York), a fenced-off white neighborhood of single-family cottages on the tip of a racially integrated Coney Island, 16 kilometers from Wall Street. Seagate requires a pass for entry.

The value of *place* is fundamental to all social life in the metropolis. Home, street and neighborhood are places which have deep and lasting meaning in most societies. The mobile U.S. metropolis is a settlement of displaced people, with automobile corridors that break up rather than unite neighborhoods, a metropolis without durable collective places. Half of all people in the U.S. change their address every five years (Barringer, 1991).

Displacement – literally the loss of place – means the loss of territory, the loss of control over land, or 'turf.' People living at the lowest levels of subsistence are the most vulnerable to displacement. Thus, displacement is a quality-of-life issue. Quality of life in the metropolis means not only the value of one's housing and community life, but stability and freedom to move voluntarily. Displacement threatens the right of people to a decent living environment in a place of their choice.

The displacement of low-income and working-class communities has been a relatively permanent feature of metropolitan development in the U.S. (see LeGates and Hartman, 1981). Its economic roots are in the real estate market. Land values in the central cities are stimulated by downtown office and luxury high-rise development, which has a ripple effect throughout the metropolitan area; low-rent houses in and near the center are then worth less than the land on which they are built. This produces pressures for removal of the people in those houses, and for the construction of more expensive housing on that land, or the conversion of the existing housing to more expensive housing (see section on *Real estate* below).

Displacement often has political as well as economic motives. The first major displacement in modern times was undertaken as a planned urban renewal project designed to break up potentially rebellious working-class neighborhoods. After the 1848 Revolution shook Paris, Georges-Eugene Haussmann, a local bureaucrat, was commissioned by Napoleon III to reshape the center of the city. Starting in 1853, Haussmann gutted the working-class neighborhoods and built monumental boulevards, like the Champs-Elysées, which allowed for quick access by troops to the center of the city. Haussmann finished his project in 1870. Despite his planned redevelopment, the Parisian working-class erupted again in 1871 and gave birth to the Paris Commune, the first working class government in history. Haussmann's urban renewal obviously did not prevent the rebellion of the Paris workers, but it may have been helpful in discouraging and then overthrowing it.

Since the rise of the metropolis in the twentieth century, the experience of Paris has been repeated in every region of the world and resulted in the involuntary removal of millions of people, including entire communities. Only war and natural disasters have displaced more people than urban renewal.

The immediate mechanisms of displacement have included: government-sponsored urban renewal and highway programs, institutional expansion, landlord abandonment, bank disinvestment and redlining, condominium and cooperative conversions, and arson (see Hartman *et al.*, 1982). In the U.S. the federal urban renewal program, mostly spurred by the 1954 Housing Act, was the single most important urban program (see Wilson, 1966; Bellush and Hausknecht, 1967). Whether displacement occurs through large-scale public intervention or gentrification (the quiet everyday functioning of the real estate market), it takes its toll on low-cost housing occupying central areas.

In the United States, displacement is racially defined. Race determines who gets displaced, how they get displaced, and what alternatives are available to them after they get displaced. African-American neighborhoods have historically been the most subject to removal by urban renewal, gentrification and housing abandonment; the federal urban renewal program became known as 'Negro removal' in the 1960s. Displacement of African-Americans usually means relocation to overcrowded ghettoes because they face suburban exclusionary practices and discrimination in employment. In contrast, many whites were 'displaced' from central cities to the suburbs during the 1950s, but it meant a general rise in their living standards and improvement in housing conditions (see Kain, 1973).

The real estate market in the United States is divided up according to a complex racial code which overlaps with income; real estate values follow racial lines as surely as they follow the principle of central location. This has also become the pattern in European metropolises with large immigrant populations from Africa and Asia, such as London and Paris. A racial division of the real estate market is also found in some developing nations, but the most dramatic example is apartheid South Africa.

The massive displacement of low-income African-American communities via urban renewal and highway projects came to an end when the federal government phased out these programs in the 1970s. Reasons for the termination of these programs include soaring central land prices, fiscal austerity and massive community protests. Now displacement occurs through the more gradual process of gentrification (Smith and Williams, 1986), which is nothing more than the 'normal' process of redevelopment through gradual land market changes without massive government intervention.

In some central cities, displacement has taken the form of large-scale housing abandonment (see Homefront, 1977). In the 1970s, New York City lost over 500,000 jobs as industry moved to the suburbs and abroad. This suppressed the portion of the local housing market serving the more stably employed workers, who followed the exodus of capital. At the same time, immigration from the South and the Caribbean supplied a new pool of surplus labor to fill lower wage jobs in the service sector. The new immigrants occupied the surplus housing stock, but because their wages were generally below the average level of subsistence they were not sufficient to support a profitable rent level for landlords. Landlords began to milk buildings for maximum rental income and depreciation tax benefits. They fell in arrears on their real estate taxes, and eventually walked away from their buildings. Many resorted to arson to collect insurance payments. As a result of this process, about 300,000 units of housing were abandoned between 1970 and 1981 (Marcuse, 1986). The City of New York eventually wound up the *de facto* owner of some 150,000 units of housing.

The costs of displacement cannot be calculated by simply adding up numbers of individuals or the values of housing, although the numbers are substantial; the costs are qualitative as well as quantitative. Displacement unties the personal and institutional knots that bind communities together, destroys the roots of civic institutions that are poorly replanted by private and public sector philanthropy, and causes extensive pain and hardship to individuals and families. It is a symptom of a wasteful regime of consumption that throws away neighborhoods, workers and families like it throws away all commodities, including land and housing, once they no longer bring in profits or if they stand in the way of the profit machine.

Central cities and the Central Business District (CBD)

The historic and universal division between central city and suburb is most pronounced in the free market U.S. model of metropolitan development. The gulf between central city ghetto and suburban affluence has been such a permanent feature throughout the post-war years that the term central city has taken on certain qualities of a myth; it has become synonomous with low-income African-American communities. This equation is only partially valid, and increasingly inaccurate as other low-income racial groups occupy the central cities, and some inner suburbs become ghettoized. Yet African-American neighborhoods are still mostly central city neighborhoods, and still have a disproportionate share of poor people.

There is one universal feature of central cities: they constitute the highest land values. Centrality is a social use value in all metropolitan areas; the center is geographically strategic because it has the greatest opportunities for access to all other parts of the metropolis. This is also true in polycentric agglomerations, where the dispersed centers are powerful. Where there is a private real estate market, the use value of centrality becomes an exchange value: land in the center is commodified and sold on the market at prices far exceeding its use value.

Minimally regulated real estate markets tend to create the Central Business District as the premiere 'neighborhood' in the metropolis, the leader in land development. The most intensive land uses, commercial and office development, bring the highest return per square foot of land and therefore tend to predominate in the highly valued central districts. Even where land is used by non-profit service institutions and public agencies, the logic of its development tends to follow the logic of the market (see Worthy, 1976). Where individual buildings and districts are assigned historic landmark status and protected from redevelopment, their preservation reinforces the market. Although the CBD in the sprawled metropolises of the post-World War II era, such as Albuquerque, New Mexico, is not as powerful as the older CBDs, and must compete with newer 'edge cities' (Garreau, 1991), it still retains special value and continues to grow.

The CBD is a daytime neighborhood whose livelihood is defined by commercial and office functions. It is only a small piece of the total land area of the metropolis – about 10 per cent – but the land values, culture and politics of the metropolis follow the lead of downtown.

The form and image of the CBD dominate urban ideology and planning in city centers. The architects for CBD buildings who leave their distinctive individual signatures on the skyline of the CBD are the most highly rewarded planners in the nation. Also recognized have been the public sector redevelopment planners like Robert Moses and Edward Logue whose bulldozers and monumental public works projects changed the face of older central cities.

Since the 1940s, the urban renewal and interstate highway programs swept away millions living in poor and working-class neighborhoods around the country in order to rebuild and provide access to the CBD. In the once industrial cores of the Eastern and Midwestern metropolitan areas, monumental civic buildings stand next to functional office towers, and the few remaining historic buildings and public spaces provide a sharp contrast to their powerful neighbors. One of the most extreme examples of the giantism of central city planning is Detroit's exclusive Renaissance Center, a towering, walled fortress of a financial center built on the rubble of workers' neighborhoods. (See Friedan and Sagalyn, 1989.)

The industrial and working-class cores of Eastern and Midwestern metropolises have been remade according to the national model of auto-centered growth whose imposition has resulted in benefits for CBDs and suburbs, and heavy social costs for central city neighborhoods. The interstate highway network, replete with spacious cloverleafs and excess condemnation, has linked the CBD and suburbs, but in the process devastated many older neighborhoods and low-cost housing.

More developed metropolitan areas have highly diverse and heterogeneous CBDs, with separate office, commercial and cultural centers. For example, New York has the Wall Street and midtown office districts, with new ones springing up in Brooklyn, Long Island City, Newark and Jersey City; some forty-eight giant shopping malls scattered throughout the region; and Manhattan's midtown entertainment district flanked by myriad suburban theaters, museums and educational institutions.

In the older metropolitan areas, central city neighborhoods tend to be more internally integrated than suburbs, both in terms of land use and social characteristics. For example, it is not uncommon to find ground-floor commercial uses below multi-family residences, or offices next to housing, in New York City and Chicago. Neighborhoods in central cities tend to be physically segregated from one another, but they also have a degree of social interaction, and the range of social strata is much greater than in the suburbs. There are more than a few examples of multi-racial and diverse central neighborhoods, whereas in the suburbs they are the exception. New York City's Lower East Side is one of the oldest mixed working-class communities in a central city. Others are (or were, thanks to displacement) Cincinnati's Over-the-Rhine, Boston's Fenway and San Francisco's Mission district. Lest we romanticize the central city, we should be reminded of the serious deficits in urban services and the quality of life in these communities, and the constant threat of displacement. Social integration is therefore possible though not inevitable in mixed use areas. It is frustrated by the centralizing tendencies of the real estate market.

Throughout the twentieth century, but especially since World War II, manufacturing fled central cities for the suburbs, regions with low labor costs, and points overseas. Large-scale manufacturers went to segregated suburban industrial parks, which drew their labor force from the segregated suburbs. In many metropolitan areas that descended directly from nineteenth-century industrial cities, however, there is still a core of small-scale industry. About 20 per cent of New York City employment, for example, is in industry. There are still many 'mixed use' neighborhoods where small-scale industry and housing are closely linked. In Brooklyn's Williamsburg and Greenpoint, many blocks have had mixtures of small industry and housing for over a century. This pattern appears to be relatively stable since with the restructuring of global production during recent decades, the fastest growing manufacturing is in small units located near major business centers where they are able to respond to flexible demand.

Is this mixture of industry and housing the integration of workplace and residence planners strive for? If it is, why have planners generally not sought to emulate or preserve these areas? Unfortunately, orthodoxy has prevailed in these cases, and planners have sought to eliminate the mixture by zoning for separate manufacturing and residential districts instead of seeking to manage this unique historical pattern in a way that would strengthen integration. Planners should try to find ways to eliminate conflicts between industry and housing, such as truck traffic on residential blocks, lack of loading docks and environmental contamination by industry, instead of trying to separate industry and housing.

One universal plague on all central city low-income communities is environmental pollution. Typically, levels of carbon monoxide and sulfur dioxide in the air are higher in central cities due to greater auto traffic and, in northern climates, the burning of heating oil. Poor people do not have air conditioners to at least filter the soot in the summertime, or second homes in the country to escape to. Surface water quality tends to be worse in central cities, and poor people lack the mobility to frequent ex-urban vacation spots. The quality of public water supplies in central cities is not necessarily any worse than in the suburbs, but poor people usually cannot afford to buy mineral water to protect themselves from the increasingly contaminated public water supply systems. In the New York metropolitan region, surface water quality correlates with location in the region and socioeconomic status. The upstream waters, which have the highest quality, are located in the suburbs, and the worst are in the central cities. The regional waste-water systems are organized so that the suburbs transmit their wastes to treatment plants located at the stems of the major drainage basins, which discharge the partially treated wastes into the polluted and often toxic surface waters near the central cities (see Angotti, 1973). And in many old industrial areas, in metropolitan regions like Chicago and New York, poor people live next door to small noxious industries in tightly packed mixed-use neighborhoods (Panos Institute, 1990).

Many myths have been advanced about the gradual integration of the U.S. metropolis by market forces. In the 1970s, popular sociology claimed that there was a new trend of middle-class whites moving 'back to the city' and blacks moving to the suburbs (see Clark, 1979; Laska and Spain, 1980). By the late 1980s, it was clear that these trends were quite limited, and reflected not a process of social integration but the displacement of jobs and housing opportunities in central cities available to low-income blacks, and the gradual extension of central city ghettoes into inner-ring suburbs. There is also the trend of growing stratification within the African-American and other 'minority' communities, and the gradual migration of middle-income blacks into gentrified portions of the central city or, more commonly, into new inner-ring segregated suburbs. In the New York metropolitan region, this is evident in pressure to integrate white-dominated Yonkers (like already-integrated nearby New Rochelle). In Yonkers, attempts to build low-income housing in white neighborhoods raised fears among whites that the municipality's ghettoized black neighborhood would expand.

In sum, the integration process in the U.S. is tenuous at best (see Taub *et al.*, 1987). This is illustrated by the 'tipping-point' phenomenon. When neighborhoods become partially integrated, most white property owners and tenants take flight, reproducing the pattern of segregation. Some claim that there is a specific point in this process at which neighborhoods 'tip' from integrated to segregated. Some have attempted to use the tipping point as a philosophy to exclude blacks. Whether or not there is a measurable tipping point, the phenomenon of resegregation is quite real. Tipping is often aided by real estate speculators – 'blockbusters' who play on racial fears to get white owners to sell cheap (Wolf, 1963).

Suburbs and the suburban metropolis

Once tiny enclaves at the periphery of cities, suburbs now comprise the majority of land and population in the metropolis (Jackson, 1985: 284). In the U.S., the majority of the metropolitan population, usually more than two-thirds, live in suburbs.

The earliest suburbs appeared before there was a metropolis, as extensions of the pre-industrial and industrial cities. The *faubourgs* in medieval Europe, for example, were spontaneous settlements by poor people, mostly peasants and traders, outside the city fortifications. The city was a container and there was only so much room in it; suburban dwellers were effectively outside the city, unprotected and subject to displacement. Successive expansions of the fortifications often incorporated them, but then they were no longer suburbs, and new suburbs, always marginal, emerged.

The first commuter suburbs integrated in the urban fabric emerged in nineteenth-century Europe and America. They were spurred by land subdivision schemes and shaped by the expansion of steam and horse railways and, in the latter half of the century, electric streetcars (Warner, 1962; Jackson, 1985).

In the twentieth century, suburban growth mushroomed with the rise of the metropolis as a qualitatively distinct settlement form. The suburb became the standard, not the exception, for new urban development. Cities expanded outward in waves of development modulated by the cyclical patterns of economic growth and the real estate market. Agricultural land was converted to urban uses (Clawson, 1971). The opening up of the 'crabgrass frontier' in the twentieth century was as much a revolutionary process as the opening up of the West in the nineteenth century (Jackson, 1985).

Perhaps the purest form of the North American metropolis is what Fishman (1987) calls the 'suburban metropolis' – the sprawled urbanized regions of the Southwest such as Los Angeles, Albuquerque and Phoenix. It is the antithesis of the 'city as container' that dominated up to the nineteenth century. It thrives on and is made possible by the seemingly limitless supply of land at the metropolitan periphery, abundant energy resources and opportunities for auto travel.

The largest wave of suburbanization took place in the post-World War II years with the massive migration of capital and labor from central city to suburbs. Industries sought land for horizontal expansion of large assembly-line and warehousing operations in the land-rich suburbs. This coincided with the hegemony of the auto–petroleum monopolies and the decline of long-distance rail and shipping industries that had favored central city industrial plants. Not surprisingly, the automobile and truck, the supreme commanders of suburban sprawl, quickly dominated the metropolitan landscape. As with the opening up of the West in the nineteenth century, this land boom was promoted and financed by government – through the Interstate Highway system, federal mortgage loan guarantees, tax benefits to homeowners, and grants to suburban municipalities for new physical and social infrastructure (see Ashton, 1978).

The latest and most recent stage of suburbanization is characterized by the growth of the communications and electronics industries, which has made possible further decentralization of industry and services following the same general patterns as the previous wave. Intra-suburban relations have begun to assume increasing proportions as the suburbs continue to grow and mature. Joel Garreau has described as 'edge cities' many of the new higher-density suburban satellites that are emerging, especially at the fringes of metropolitan areas that have relatively undeveloped central cities (Garreau, 1991).

In analyzing industrial flight from central cities, David Gordon (1978) emphasizes the importance to industry of escaping the restrictions imposed by organized labor and gaining greater control over the workforce. This was certainly a factor in suburbanization, as it is in every decision by capital to relocate. However, Gordon's analysis is conspiratorial and ignores the multiple dimensions of decisionmaking by firms. It ignores the very real changes in industrial technology, the international market, and the labor force itself. By the time industry made its massive move to the suburbs, the industrial labor movement, devastated by McCarthyism and little concerned about the plight of central cities, was willingly coopted into the new suburban-based labor aristocracy. And as revealed by the recent widely publicized strike by Local P-9 of the United Food and Commercial Workers against the Hormel Meatpacking Company in the Minnesota town of Austin (population 21,000), industry did not necessarily escape either labor unions or labor militancy by moving to towns and small cities, to the suburbs, or from the Snowbelt to the Sunbelt.

Suburban development accompanied an overall improvement in the standard of living of the U.S. working class, even though differences within the class widened. It was made possible by the internationalization of capital in the post-war years, which produced an unprecedented accumulation of surplus. It was also made possible by an increase in the actual and potential strength of a large privileged stratum of labor, which sought, and received, a share of the abundance.

The growth of suburbs coincided with the revolution in consumption relations in the twentieth century as much as the revolution in production relations. As Thorsten Veblen noted early on, this was to be the century of conspicuous consumption (Veblen, 1961). The suburb has been the haven of conspicuous consumption, where the status seekers, men in white collars and grey flannel suits, erected their individual castles. Suburbia is a complex and highly stratified social landscape, as many sociological studies have demonstrated, and the stratification extends to all levels of the working class, including blue collar and white collar (see Dobriner, 1963; Masotti and Hadden, 1973; Baldassare, 1986).

The hierarchy of suburbs is a function of the degree of exclusion and segregation. The more a community excludes, the more it is worth to its residents, both in terms of property values and social values. It is considered a virtue to have to drive miles to a regional shopping mall or highway commercial strip – in someone else's back yard – but painful to have to walk to a local grocery store and encounter people on the way, especially people with a different skin color. It is considered a virtue to commute to work a couple of hours each day rather than live near a business or manufacturing district. This is made more comfortable by the tape deck and radio in the car, where the isolation of the dream house is drawn into public spaces. The American dream is a suburban homesteader's dream of sweet alienation and exclusion.

Suburbanization entered a new phase in the 1970s. Differentiation among suburbs grew as central city ghettoes expanded into the inner-ring suburbs, high-tech and central office functions were dispersed to the periphery of the metropolis, and redevelopment pressures struck once-exclusive blue blood enclaves. Despite the greater divisions, the basic central city/suburban dichotomy remains. Even though the mythical nuclear family has rapidly lost its hold – today only one-fourth of all households fit the stereotype – the American dream house on a separate lot is still the ideal.

Contrary to the view that there is a plot to disperse minorities and poor people to the suburbs, thereby diluting their political power, the government continues to uphold the rights of suburban exclusion above the civil rights of minorities. The trend is not toward the 'deconcentration' of minority neighborhoods, but rather displacement and resegregation in new ghettoes of the central cities and proximate inner suburbs. Fair housing legislation has had little impact in the suburbs, and the government programs to build affordable low-cost housing in the suburbs have been minimally funded and often unavailable to minorities. (See Champion, 1989.)

The central place in the suburbs is the auto-accessible shopping mall, not the main street of old or the central city CBD. The shopping mall is in many ways the suburban CBD. Shopping malls and highway commercial strips were planned to be separate from housing and have contributed to the excessive need for auto use. They consume vast areas of land for their buildings and parking. Mall development contributed to the decline of many older central shopping districts and has contributed to the degradation of environmental quality.

In recent years, suburban planners have begun to retrofit shopping centers and transform them in the image of the CBD. Social and cultural functions and recreation are now prominent, and often commercialized. Office space – lacking in the first shopping centers – now appears in suburban malls. As new waves of land value increases encircle some shopping centers, more densely configured housing is appearing and some malls are beginning to take on the appearance of a CBD with a huge parking plaza. Many suburban malls were not designed to afford the variety and level of functions available in central cities, but the growing demands for suburban expansion have given rise to new super-malls and self-contained 'edge cities' (Garreau, 1991).

Another trend is the introduction of planned unit developments (PUDs) and cluster zoning, devices that seek to concentrate suburban building and leave more land undeveloped. These techniques allow for greater efficiencies within large tract subdivisions and avoid the monotonous serial reproduction of tract developments like the classical Levittown. However, they usually do not alter the overall density, which remains relatively low, and only distribute it differently (see American Society of Planning Officials, 1960).

Government intervention in suburban development has sought to rationalize the expansion of production, circulation and transportation systems by introducing elements of metropolitan-wide planning. However, government runs up against the exclusionary logic upon which the metropolitan spatial pattern is founded. The social compact upon which suburban municipalities were established provides them with political autonomy and control, within limits established by the market, over land use. Municipal governments were founded on the mercantile principle of competition for tax ratables and self-financing of services, which means 'clean' industry and exclusion of the poor. Patrick Ashton put the contradiction well:

> suburbs are part and parcel of the logic of capitalist development. On the one hand, they are a spatial manifestation of the many social divisions created by capitalist society. On the other hand, they have resulted from the differential ability of various groups to organize themselves to protect competitive advantages . . . given the logic of competition within capitalism, those groups using suburbs to protect certain advantages are forced to jealously guard these privileges. They cannot afford to abdicate clear-cut short-run advantage for more ambiguous long-run gains. Thus metropolis-wide transportation and land use planning is extremely difficult, if not impossible, to implement. Yet without it the continued profitability and stability of the metropolis and all of capitalist society is threatened.
>
> (Ashton, 1978: 82–3)

The auto-centered metropolis

The social life and physical form of the U.S. metropolis owe a particular debt to the automobile, and the automobile would not be the popular commodity it is today if it did not dominate transportation in the metropolis. Just as trade routes shaped the ancient settlements, ports became urban cores, and the railroad opened up the West, the highway has become synonymous with the North American metropolitan way of life in the twentieth century.

The high level of residential, employment and transportation mobility in the U.S. is in great measure made possible by the automobile. People can change homes without changing jobs, change jobs without changing homes, and have a wide range of choices for recreational trips. Mobility is not just a result of technology, but is a central element in the U.S. regime of accumulation and urbanization.

Capital and labor are highly mobile, and there is a premium on quick profits and short-term earnings.

The dominance of the auto and petroleum industries in the U.S. economy made possible and perpetuates the auto-centered model of metropolitan growth. Five of the top ten corporations in the United States are auto makers or oil producers. The U.S. economy depends on expansion of these industries, even as they are being squeezed by foreign competitors. (See Ridgeway, 1970; Weisberg, 1971; Tanzer, 1974.)

According to Daniel Yergin:

> Ours has become a 'Hydrocarbon Society.' . . . In the twentieth century, oil, supplemented by natural gas, toppled King Coal from his throne as the power source for the industrial world. Oil also became the basis of the great postwar suburbanization movement. . . . It is oil that makes possible where we live, how we live, how we commute to work, how we travel – even where we conduct our courtships. It is the lifeblood of suburban communities. . . . The twentieth century rightly deserves the title 'the century of oil.'
>
> (Yergin, 1991: 13–15)

Not to be overshadowed by oil advocates, leading experts on the auto industry claim, 'The world stands at the centenary of the automotive age' (Altschuler *et al.*, 1984: 1). Advocates of oil and autos both have a point. The development of oil and the auto – and we must add the metropolis – have distinguished this century from all previous ones. But in the international competition between the United States, Europe and Japan, the most strategic industry is auto and in the long run substitutes for petroleum may well prove both feasible and necessary.

The auto-centered model of metropolitan development provides many opportunities for mobility, though the opportunities are distributed unequally. The auto is one of the most flexible modes of long-distance transportation for multiple passengers, and one of the most suited for recreational trips. However, where it has been adopted on a mass scale as the main means of transport, it degrades the overall quality of urban life, produces widespread pollution and contributes significantly to the global environmental crisis.

Los Angeles and the metropolitan areas of the West – the suburban metropolises – were built around the logic of the automobile, despite some brief flirtations with mass transit. They may be contrasted with the older metropolitan areas of the Eastern and Central states, like New York, where the suburban/auto model was imposed on industrial cores with developed neighborhoods and mass transit systems. The result there has been transportation chaos. New York, the central city with the largest mass transit system in the country, has some of the worst traffic problems. New York has a dual system: on the one hand mass transit which disproportionately serves the poor and has inferior service levels and declining ridership; and on the other hand the vast street grid, which serves higher income auto users, and is growing in use despite the problems of congestion and pollution.

There are now some 400–500 million automobiles in the world and they are responsible for over half of all petroleum consumption. The United States, home of Ford and Fordism, has more cars per capita than any other major consuming nation. The U.S. has a wasteful oil-based regime of accumulation which, despite challenges from the more energy-efficient regimes of Europe and Japan, it continues to impose around the world through economic blackmail and military force. In the U.S. there is one car for every two people. While it accounts for only 20 per cent of world auto production (Europe and Japan produce two-thirds of the total), the U.S. remains the largest consumer of cars (Altschuler *et al.*, 1984: 108).

Wherever it predominates the automobile–petroleum combination has wreaked havoc on metropolitan areas. In older European and Third World metros, the auto has been allowed to take over and destroy public plazas and narrow streets designed for non-motorized vehicles, and sidewalks designed for pedestrians. Planners have let the auto take over many private dwelling spaces built for human habitation. The result is the abandonment of public spaces and retreat to private homes by people. The conversion of public spaces to parking lots and roadways brings with it greater threats to public safety, high pollution levels and greater street maintenance costs.

In the U.S., where public spaces never achieved the status they did in other cultures, the automobile has been a symbol of an egotistic and wasteful mode of individual consumption likewise nurtured by the single-family dream house, television and home video center. Mass advertising has consecrated the automobile as the icon of personal ambition and male aggressiveness.

The auto-centered metropolis wastes land. About 25–30 per cent of its land is dedicated to the movement and storage of the four-wheeled machine. Reliance on the auto as the main means of transportation makes possible the sprawl that eats up ever greater portions of the suburban periphery.

The auto-centered metropolis is *less accessible* to the majority of people than the metropolis that favors mass transit and walking. Mass transit modes have higher capacities per passenger mile and in practice achieve average speeds of 20–40 miles per hour in central cities while the auto averages 7–10 mph. The classical orthodoxy of traffic engineering that increasing road capacity lowers traffic congestion is belied by the fact that metropolitan areas with serious congestion problems, like Los Angeles, have expansive freeways. Where mass transit is prolific and subsidized, mostly in metropolitan areas outside the U.S., it benefits people of all income levels who do not own cars and serves as the main means of transport, especially for work trips (see Smerk, 1968).

The auto is also *less energy-efficient* than mass transit. The auto consumes at least twice as much energy per 1,000 passenger miles as the subway and streetcar (see Meyer, Kain and Wohl, 1965; Smerk, 1968; Lowe, 1990). It is perhaps the major contributor to global air pollution and the greenhouse effect. It is responsible for 60–80 per cent of the carbon monoxide and particulate matter in the air and is one of the major contributors to heart and lung diseases, especially cancer. In 1988, the U.S. discharged 4,804 metric tons of carbon dioxide into the atmosphere, more than any other country, the second highest rate per capita after East Germany (Stevens, 1991).

The auto is the most dangerous mode of transportation to its drivers and passengers as well as to those who happen to be in the way of drivers. Each year, approximately 45,000 die from auto accidents in the United States alone. In New York City, there is an average of one pedestrian fatality a day involving a motor vehicle. In the auto-accessible metropolis, children play in the street at risk of their lives. Cyclists must also beware. Services are not within walking distance but strung along commercial highway strips and shopping centers surrounded by huge parking lots. Services once available, monitored and controlled close to home are now auto-based, mobile and less effective. For example, when the cop on the beat in central city neighborhoods was replaced by the squad car on call, the quality of services went down and crime went up. When streets are cleaned and garbage collected by anonymous truckers, the streets stay dirty because no one attends to the many scraps and spaces that need attending in between pickups and off the routes. Streets are now caked with crank-case oil, and road salts in the winter, which runoff into the storm-water system and foul waste-treatment plants and waterways. Doctors and health-care workers are not in communities but in cars going from their homes to mammoth centralized health-care facilities, and for many the ambulance and emergency room is the only substitute for home health care. Schools are reached by car or school bus, not on foot.

Given the dependency of the U.S. economy on the auto and petroleum industries, and the strength of the auto lobby in Washington, it is no wonder that the federal government has resisted greater subsidies to mass transit, and been reluctant to go along with international accords to stop global warming. Perhaps the largest public works project ever undertaken by any government in the world is the federal Interstate Highway system, a network of some 50,000 miles of asphalt and concrete first conceived in 1944 as the Defense Highway System. The powerful highway lobby, consisting of groups representing auto manufacturers, oil companies, building contractors and car owners, has managed to protect the Highway Trust Fund as the self-perpetuating source of revenue for the interstate system, although the 1991 Transportation Act chipped away some of this protection. The Trust Fund comes from gasoline taxes, and almost all of it goes back into road construction (see Leavitt, 1970).

The orthodox auto-minded planners of the U.S. continue to promote the myth that air pollution can be diminished by tightening up emission controls, when it is clear that those countries that have the tightest emission controls (the U.S. in particular) are the biggest polluters because they use regulations to rationalize increased auto use, and pollution, not to decrease them. The dominant approach to pollution is summed up in the term 'risk management,' which means engineers try to calculate an 'acceptable' or optimum level of human and environmental damage to protect the high levels of overall consumption.

Auto advocates have promoted the image of the auto-centered metropolis as a model of free choice, where people are free to choose the transportation they want. The problem is that the auto–petroleum lobby uses all its economic and political power in Washington to ensure that mass transit does not become a real choice for most Americans (Leavitt, 1970). Between the 1930s and 1950s, corporations dominated by the auto, petroleum and tire industries bought up the extensive streetcar systems in most U.S. metropolises and put them out of business to make room for Detroit's cars, buses and trucks (Snell, 1973).

The hidden subsidies to autos are substantial and auto users pay but a fraction – perhaps one-fifth – of the social costs of their travel when they fill up their gas tanks (Smerk, 1968; 15, 83–4, 211; Lowe, 1990). Thus, gasoline prices and taxes do not cover many roadway maintenance costs, policing, traffic control or snow removal costs. Auto users do not cover the costs of traffic fatalities and injuries, or sickness and death due to air pollution, or the military 'protection' provided oil companies in the Middle East (see Lowe, 1990). Reliance on the automobile as the main mode of transportation has contributed to the sprawled suburban development pattern and the associated inefficiencies in land use and infrastructure. Many city planners consider the greatest costs of the tyranny of the auto to be the loss of community, displacement of people, and the degradation in the quality of life.

We shall return to this theme, and alternative planning strategies, in the last chapter. Now we turn to summarize the theoretical principles underlying the growth and planning of the U.S. metropolis, particularly as they affect the key issues of quality of life and displacement. We have shown in Chapter 1 how the regime of accumulation is a fundamental determinant of urbanization and planning. Now we turn to three other factors: the real estate market, profit and labor force reproduction. The real estate market and profit are closely related to the dynamics of displacement, and the regimes of accumulation and reproduction are bound up with the quality of life.

Real estate, housing and displacement

In the U.S. metropolis, the real estate market plays a particularly prominent role in the allocation of land use. Comprehensive planning plays a minimal role. The main regulatory tool is zoning, which tends to follow and codify land-use changes rather than determine them. While a powerful force in metropolitan development, the real estate market is not a 'free' market, because it is shaped by many local and national policies. These policies favor the dominant centralizing tendencies of the market, and do not add up to a coherent approach to land-use planning.

Economic analyses of internal metropolitan growth most often leave out the determining role of the real estate market. Greater determination is often attributed to the international movement of capital, or the role of enterprises and consumption patterns. Clearly, broader economic and political factors provide a context in which the metropolis grows, but cannot alone explain its growth (see Bandyopadhyay, 1982).

The real estate market is the principal mechanism for the allocation of land use in the U.S metropolis. In other words, the exchange value of land, and not its use value, prevails (Harvey, 1973; Logan and Molotch, 1987). Central locations yield higher land and building rents. High rent, high-rise commercial space and luxury housing cram the central districts while low rent, low-density uses are found in the periphery.

Land values increase at the center and have a ripple effect as they spread to surrounding areas. As the waves spread outward, there may be a temporary depression in land values like an ebb tide, resulting in disinvestment and housing abandonment, but ultimately values in the periphery rise in response to a rise in central values. Always some individual buildings and lots are excepted from the rule due to the influence of public institutions, historic districts and traditions, and geographic anomalies. (See Hurd, 1903; Hoover, 1968; Harvey, 1973; Feagin, 1986.)

The real estate market is shaped by government investment in infrastructure, tax policy, market incentives and land-use regulations. However, in the United States, where local governments tend to reflect real estate interests, land regulation is molded to meet the common objectives of property owners and developers. Regulations constitute the minimum rules of the game that mediate conflicts among property owners, and between property owners and other social groups and institutions.

Since the generalized commodification of urban land, the expansion of real estate values has been bound up with the displacement of low-income and working-class neighborhoods:

> The expansion of the big modern cities gives the land in certain sections of them, particularly in those which are centrally situated, an artificial and often enormously increasing value; the buildings erected in these areas depress this value, instead of increasing it, because they no longer correspond to the changed circumstances. They are pulled down and replaced by others. This takes place above all with centrally located workers' houses, whose rents, even with the greatest overcrowding, can never, or only very slowly, increase above a certain maximum.
> (Engels, 1975: 20)

Land markets are intertwined with, but not identical to, housing markets. Also, displacement is bound up with, but not the cause of, the shortage of low-income housing. Even when there is minimal displacement, there may be severe housing shortages, which is the case in many European countries with significantly less mobility of capital and labor.

What is the source of housing shortages? Orthodox 'trickle-down' theory reduces the problem to one of short supply, and the solution to building more housing units. The less orthodox acknowledge that in a free market system there is always a housing shortage of one kind or another, especially for low-income people. In the U.S., the housing shortage has always affected low-income people the most. Public subsidies for middle-income homeowners exceed housing subsidies for low-income people by a ratio of five to one (Dolbeare, 1986). Low subsidy levels for low-income housing, and low incomes, contribute to the chronic shortage of housing in a housing market that is hardly 'free' (see Dolbeare, 1983).

Housing supply is always plentiful in areas of increasing land values, and among higher income strata of the population. The market can easily respond to changing needs among higher-income people, and often produces a surplus at the top of the market. These surpluses typically lead to temporary vacancies rather than a 'trickle-down' to lower income groups. (For an excellent critique of orthodox theory, see Gilderbloom and Appelbaum, 1988.)

Because of the highly stratified labor market, lower income individuals and households can only afford rental housing. No bank will provide them with a mortgage because they do not have sufficient income to pay it off, or the long-term job stability demanded by financing institutions. And when there is no room in the private housing market, the only alternatives left are public housing or homelessness (in its various forms, including overcrowding and doubling up).

Because of the durability of racial stratification in the U.S., publicly financed low-income housing has been one of the few alternatives available to low-income communities, both to mitigate displacement and to guarantee a minimal housing supply in an exclusionary market. But public housing by itself cannot resolve, and obviously has not resolved, the underlying economic problem of inequality.

Racial exclusion has been at the center of the public housing debate in the United States, and partially explains the limited supply of public housing. Many local housing authorities faced early opposition to the dispersal of subsidized housing units throughout central cities, leaving them with the sole option of building high-rise ghettoes in segregated neighborhoods (see Meyerson and Banfield, 1955). Opponents of public housing in Congress insured they would not have adequate funds for maintenance and services, and then could point to the poor maintenance to demonstrate public housing's failure. In the 1970s, the Nixon Administration proposed scattered-site housing in the suburbs, but this was bound to fail given the political strength of suburbs – an outcome the Nixon Administration, ever anxious to cut public housing subsidies, was certainly able to live with.

It is significant that the appearance of several million homeless people on the streets of central cities in the 1980s paralleled the virtual halt in construction of publicly subsidized housing under the Reagan Administration. At the same time federal housing assistance to new construction dried up, gentrification removed many rental units from the potential low-cost housing stock. Some of these units became condominiums and cooperatives whose cost was out of the range of low-income people. In addition, privatization of public housing became a top priority for conservative administrations, and considerable progress toward this end was made in Thatcher's England (Hamnett and Randolph, 1988). In sum, the austerity policies of the 1980s have sought to remove the pillars of the liberal welfare state and facilitate a new kind of primitive accumulation based on very low levels of subsistence.

Profit

If the real estate market is the determining economic force in land use, the profit principle is the driving force underlying the entire economy: profit in production, consumption and land. In the U.S. regime of accumulation there is a premium on quick profits – the Gold Rush spirit of venture capitalism. Inventiveness entails the ability to turn anything into a commodity, including land, housing, public places, information, services and neighborhoods. The approach to land use in the U.S. is imbued with the spirit of nineteenth-century 'pioneers,' backed by the railroad companies and government, who displaced Native Americans from the vast territories of the West and turned their land to speculative use.

Throughout the metropolis, property owners and financial institutions seek to maximize their profits, and meet social needs only insofar as they contribute to the protection and enhancement of their property rights and profit margins. Parks and playgrounds and other social amenities that are usually publicly owned and operated do not directly produce profits to property owners. However, their location is a function of the value of adjacent property, which they invariably enhance. It is no accident, for example, that New York's Central Park boosts property values (though unevenly) on its edges. Other public facilities, such as solid waste transfer stations, drug treatment centers or homeless shelters, that lower adjacent property values, tend to be excluded by the organized power of property owners and located in poor neighborhoods without the political power to exclude or control them.

Owners of real estate make up only one sector of profit-maximizers in the metropolis. They are characteristically fractured and divided into many localized interests, and there is usually no little friction between large and small landholders, central city and suburban owners, commercial and residential owners, institutional and individual owners, etc. Other groups of profit-maximizers are in the production sector. They usually have more of a national and international base but in many metropolitan areas include large numbers of small entrepreneurs dependent on monopoly-related contracts and investment patterns. The monopoly capitalists, in production and finance, are mostly loyal to one another in local politics, and natural allies of (if not merged with) Central Business District real estate interests. However, the banks and lending institutions that as a group finance almost all property acquisition and development play a significantly more active role than other businesses, especially in their favorite neighborhood, the CBD.

Finally, the metropolis has given us a new breed of entrepreneur whose profit depends on urban services. The uniquely metropolitan-based capitalists contract with government for waste disposal, bus service and public works construction, for example. They also include, and are often dominated by, state-sanctioned private utilities and authorities, and personal service enterprises whose profits depend primarily on the general level of consumption in the neighborhoods of the metropolis.

All of these sectors constitute the leadership of the metropolitan pro-growth blocs that generally dominate local government (see Mollenkopf, 1983). Headed by financial and monopoly interests, municipal and special authority bondholders, and Central Business District real estate, and allied with the auto–petroleum bloc at the national level, their unifying goal is growth, their common interest the protection of private property. They may compete among one another for a share of the profits generated by growth, but they cooperate to protect the institutions of private property and guarantee state support for them. They include in their political blocs sectors of the organized working class, particularly the well-heeled construction unions and government employee associations who stand to get jobs and housing from new development. But labor's interest is usually not direct profit on land or capital (though their pension funds may profit), and labor's role is usually subordinate.

The contradictions within these blocs are many. Entrepreneurs in production and finance want to keep the costs of labor reproduction low, and therefore seek a more efficient metropolis. Entrepreneurs in services and organized labor want greater expenditures, both public and private, for the metropolitan infrastructure. Real estate interests want more state spending to improve their property values, but without increasing local taxes. Local politics reveals all the contradictions within and among these sectors, added to which are the myriad interests of labor, communities and other elements of civil society. Their common objective, however, is growth. Indeed, as Logan and Molotch suggest (1987: 50–98), the metropolis is perceived as a 'growth machine.'

We shall return to pro-growth blocs in Chapter 6.

Unequal reproduction and consumption

At a more general level, the inner dynamic of the U.S. metropolis is bound up with the economic laws of labor-force reproduction. In the U.S. these laws operate to reinforce inequalities in the levels of consumption.

The physical and social infrastructure of the metropolis – housing, roads, water supply, sewers, schools, hospitals – are part of the conditions required to maintain and replenish capital's most essential commodity, labor-power. The quality of urban life is a function of the minimum that capital has to spend for the reproduction of labor, especially reproduction extended over a lengthy period of time and many cycles of capital reproduction. The cost of the extended reproduction of labor is a historically developed factor determined by the overall level of productivity and accumulation of surplus, and the relative political strength of labor and capital.

David Harvey has argued that much of the metropolitan infrastructure has become part of fixed capital, and its growth has contributed to the crisis of capital. Folin (1981) has shown instead how the infrastructure is essentially part of the total economic surplus and a function of income consumption.

The differences in the quality of urban life between Mexico City and Los Angeles allow for the reproduction of Mexican labor in Mexico at a lower cost. This provides a vast labor reserve for transnational capital, which they can use either in Mexico City or Los Angeles as global market conditions change. If the standard of living is too low because the social and urban infrastructure is inadequately developed, labor may not have the requisite education and skills needed to guarantee the quality of production. If the standard of living is too high, labor costs will not be competitive and investors will go elsewhere. At present, U.S. investors do not find it profitable to put more money in production in Mexico City because the standard of living there, and thus the value of labor, is too low to be competitive with Mexican (and other) labor located in the slums of Los Angeles. However, many U.S.-based corporations find it profitable to invest in Mexico City, or the *maquiladoras* of Laredo and Matamoros (where management can cross the border and live at a higher standard of living), instead of Los Angeles.

In this sense, metropolitan inequalities do not simply reflect differences between classes, but mainly differences *among nations* and *within classes*. Differences of national origin, race, gender, social status, sexual preference and physical abilities mediate class differences and are usually more important in the social and political life of the metropolis than strictly economic or income differences.

The capitalist accumulation process in the twentieth century has produced an international division of labor and division of space in which the poorest sectors of labor live separately, below the average level of subsistence, many below the minimum level of subsistence. An underdeveloped metropolitan environment in the developing nations, and in poor neighborhoods of the developed nations, keeps wage levels low in the poor areas, suppresses overall wage levels and increases the ability of capital to control labor. Inequality has a spatial dimension and the metropolis occupies that space.

Stratification of the labor market is the basic condition for exclusion and segregation of communities. Labor is stratified in all economic systems, but the U.S. regime of accumulation requires a particular stratum that makes segregation obligatory – the reserve army of labor. To have a ready reserve of low-paid labor, capital requires reserve neighborhoods. If the labor reserve in any metropolis were to be integrated within all communities, their standard of living would not be sufficiently different from the rest of labor and they could not function as a reserve army. Profit levels throughout the metropolis would be lowered because firms would have to pay higher wages overall and would compete unfavorably with employers in other metropolitan areas that could take advantage of labor reserves and the lower general wage levels that go along with them.

Significant sectors of capital find a high degree of stratification and racial segregation to be unnecessary because they do not rely on lower-paid labor. They are concerned about the social and political consequences of extreme racial and class polarization, and support progressive social legislation. However, in a country born in slavery, racism has taken on a life of its own and the roots run deeper than economics.

A socially and spatially integrated metropolis could be, at least in theory, a more efficient metropolis for capital. There would be maximum mobility for labor, lower transport costs, and the per capita cost of reproducing the labor force would be lower. However, when labor is a commodity, individual firms compete for labor. This means there are substantial wage differentials, and these in turn translate into substantially different levels of disposable income. Thus, overall metropolitan efficiency is frustrated by the commodification of labor.

Different income levels articulate the metropolitan property market, which distributes households in space. The real estate industry seeks to rent or sell to people with the highest incomes. To do so, property owners hope to imprint a special address on their holdings. Exclusiveness raises the value of property, whether it is in the more expensive central locations or at the suburban fringe, and secures profit levels for sellers. Parts of the real estate industry also make profits from low-valued property without exclusive addresses, but their profit depends on somewhat riskier, though no less profitable, investment strategies in low-income segregated areas (Sternlieb, 1966). The slumlord regime of profit-making is thus a complement to the exclusive enclave regime of profit-making.

Because subsistence and consumption have become increasingly collectivized in the modern metropolis, and residential mobility is high, one could theoretically change one's standard of living by changing one's address. In other words, poor people would just move into better neighborhoods. This was the fantastic proposition underlying the conservative proposals in the United States for housing vouchers to replace public housing subsidies. However, this kind of instant integration is blocked by racial discrimination, differences in wage levels and disposable incomes, and differences in rent and housing prices.

Part of the degraded level of consumption of poor people is vulnerability to displacement. The residential, employment and transportation mobility that are assets for the majority of the metropolitan population are liabilities for the poor. Low-income and minority communities are more likely to be displaced by urban renewal and public works projects, institutional or business district expansion, and the rising tide of central land values.

Differences in consumption levels, and property values, are no less significant in the dependent metropolis, where often the majority live at what appear to be below-subsistence levels in the so-called informal sector. As stratification of the labor force becomes more pronounced, poverty and the informal sector tend to grow, and with this growth metropolitan inequalities are reproduced as local real estate investors reap new benefits (see Mingione, 1983). This is one reason for the alliances between industrial and real estate capital. In some cases, differences in consumption levels may exceed diffences in income, producing wider metropolitan inequalities (Saunders, 1986).

The metropolitan environment is the medium in which labor collectively consumes the portion of the surplus allocated for its extended reproduction. It is the environment of *collective consumption*. The infrastructure of services in the metropolis is in effect a collective wage, a contribution from the total social surplus that establishes the standard of living of workers. Instead of spending this wage on individual consumption, workers consume public goods; that is, they consume collectively (see Castells, 1977).

While subsistence and consumption have become more and more collectivized in the metropolis, the collectivization process has been unequal. Racial and ethnic barriers prevent parity of subsistence levels, and reinforce the structure of economic exclusion.

Collective consumption may lead, in the long run, to greater social equality because it presumes a wider availability of goods and services. It may make possible greater social and physical planning. It may be one of the economic building blocks of the transition to socialism. However, it has so far failed to bridge the gap between ghetto and gold coast.

There have been powerful countervailing trends to the collectivization of consumption. In the Reagan–Bush era, privatization of public goods, such as public housing and education, have become a mainstay of the conservative agenda. The privatization drive is but the latest manifestation of commodity fetishism – an impulse to turn everything into a commodity for distribution on the market, including metropolitan services, places and values. So far, privatization has yielded minimal results, and the 1980s' deregulation hysteria has been followed by quiet government bailouts and re-regulation. The Reaganite ideology of nineteenth-century *laissez-faire* capitalism, though ideologically triumphant, has been unable to reverse the epochal trends in political economy, though it may have slowed them temporarily. Indeed, Reaganites could learn a lesson from the failure of ideology in the former Soviet Union.

Private property in the means of production, the foundation of capitalism, does not require private property in consumption, and may at times benefit from collective consumption. Indeed, as the economic rise of Europe over America suggests, capitalism in its mature stage is not only able to survive with a collectivized form of consumption, but apparently requires it to maintain a high rate of accumulation.

Similarly, social property in production, the foundation of socialism, does not require collective consumption, and may benefit from private consumption. Failure to understand this was part of the crisis in Eastern Europe and the former Soviet Union. In any case, in the broadest sweep of history the gap between regime of production and regime of consumption is narrowing and, as more economies become mixed, with different ownership forms, the distinctions between ownership and control of production and consumption are bound to narrow even further. In sum, all spheres of capital, including consumption, are becoming more closely tied to production. Thus,

> While circulation appeared at first as a constant magnitude, it here appears as a moving magnitude, being expanded by production itself . . . as an essentially all-embracing presupposition and moment of production itself.
>
> (Marx, 1973: 407–8)

This view differs from that of Manuel Castells (1977) and Peter Saunders (1986), who tend to see the spheres of production and consumption as structurally separate. Their dualist approach leads them to see political struggles over production as qualitatively distinct from consumption issues. In the modern metropolis, however, these two spheres are becoming increasingly intertwined and inseparable.

Conclusion

In conclusion, the U.S. metropolis is an unequal but highly mobile metropolis.

Metropolitan development, like all social processes, is necessarily uneven. Metropolitan areas develop first in some parts of the world, and some develop more quickly than others. However, *inequality* between and within metropolitan areas is not inevitable but a structural characteristic of a particular type of capitalist development. It reflects fundamental social and economic inequalities in the mode of production and regime of accumulation (see N. Smith, 1984; 1986).

The unequal market-driven metropolis is an egotistic one, designed for individual gratification over social welfare, private over public good, commercial over human interaction, electronic over direct communication, property over people, exclusion over integration, and consumption over culture. It consumes vast amounts of land, water, food, oil and gas and pollutes air, soil and water as if all resources were in limitless supply, as they seemed to be for a long time in North America.

Planners need a different vision of the metropolitan community based on social as well as individual welfare, public as well as private good, human as well as commercial interaction, direct as well as electronic communication, people and not property, integration and not exclusion and culture before consumption. We can learn how to do this by understanding the spirit of real communities as they now exist, their public and private places, their daily human problems and interactions; and by expanding these places until they dominate the metropolitan landscape and are linked with private places. We cannot do this by applying pre-conceived notions and models of the ideal metropolitan environment. We need to go to the street corners, parks, sidewalks, storefronts, hallways, theaters, bus stops, subways, and every place where people seek to expand their lives, and join these livelihoods with the neighborhood and metropolitan experience.

In the long run, a better environment is consistent with the reduction of metropolitan inequalities. For the time being, however, the benefits of technological advances and pollution reduction are bound to accrue to the developed countries, and the gap between them and the developing nations will grow.

The accessible metropolis will favor equitable solutions to the transportation problem, have a variety of opportunities for mobility, and encourage mass transit and pedestrian circulation (Schaeffer and

Sclar, 1975). Circulation will be planned so that all people have maximum mobility, not just those who drive automobiles. Transportation will be a means of bringing people together in a community instead of driving them (literally) apart. As in Jane Jacobs' diverse community, there will be people and eyes on the street, while cars will be controlled instead of controlling (Jacobs, 1961). Pedestrians and people in face-to-face interaction will be catered to, not cars. Public transit will be subsidized and perhaps free, especially for children and the elderly. Alternative modes of transportation, like bicycles and jitneys, also have their place. This metropolis and community is achievable, and elements of it have begun to appear (though there are counter-trends) in places like Russia, China and Cuba, and in metropolitan areas with auto-free transportation experiments in Germany, the Netherlands, Italy and other European countries. But this model of community development poses a direct challenge to the hegemony of the auto and petroleum monopolies, and their pro-growth alliance.

We shall explore these prospects further in the final chapter. Now let us turn to the dependent metropolis.

3

THE DEPENDENT METROPOLIS

Development and inequality

> If every city is like a chess game, the day that I learn the rules I will finally command my empire, even if I never know all the cities within it.
>
> Kublai Khan speaking to Marco Polo in *Le Città Invisibili*
> by Italo Calvino.

Since North America is the only region where the majority of people live in metropolitan areas, its metropolis is often thought to be a universal prototype, with characteristics fundamental to the metropolis everywhere (see Hoselitz, 1969). But this is not the case. The North American metropolis is in many ways atypical, and has many characteristics rarely found in other parts of the world.

As with all other social phenomena, the metropolis does not develop uniformly, in the same way in every part of the world. Some parts of the world develop metropolitan areas first, and these may mature sooner than others. However, the most mature metropolis at any given time is not necessarily a 'model' for the future, or a picture of what future metropolitan areas will look like – no more than nineteenth-century Europe was a model for and prefiguration of twentieth-century Africa or Asia.

THE DEPENDENT METROPOLIS: AN OVERVIEW

The most prevalent metropolis today is the metropolis of Africa, Asia and Latin America. The term *dependent metropolis* refers here to metropolitan areas whose economies depend heavily on the developed capitalist countries, and which usually rely on export industries. It does not necessarily imply any dependent urban structure, i.e., the dependency of one metropolis on another, as in some versions of 'dependency theory.' Nor does it imply a single world system of dependent and dominant settlements, as in 'world systems theory' (Wallerstein, 1974). It refers to the urban form of dependent economic structures. These forms differ substantially from one nation to another, just as there are vastly different degrees of economic dependency. Furthermore, the term is not used in a static sense as an ideal type, but as a dynamic category based on existing and changing economic and social relations.

The differences between the dependent metropolis and the northern developed metropolis are many. In the dependent metropolis about two-thirds of the population live in poverty, many in shanties without water, sewerage or basic human services. By way of contrast, about one-third of the population of the U.S. metropolis lives in poverty, though by Third World standards their shelter and services tend to be more substantial than in the Third World. The dependent metropolis stands apart from and dominates an imbalanced rural settlement structure. The U.S. metropolis is located within a balanced, metropolitanized national settlement structure. The dependent metropolis is among the largest and fastest growing in the world. The U.S. metropolis is relatively stable in size.

In North America there are only a few extremely large metropolitan areas, like New York and Los Angeles. On the other hand, twelve of the twenty largest metropolises in the world are in developing countries, and they are growing at a faster rate than those of Europe and North America grew in the early twentieth century, mostly due to the higher rate of natural increase in population in developing countries (Preston, 1988: 14).

The dependent metropolis is the subordinate element in an unequal international settlement structure, and it reproduces the unequal relations that lie at the heart of the international division of labor. It evolved from the colonial city and bears the marks of colonial dependence. The dependent metropolis is an integral part of the neocolonial system of dependent capitalism, in which relations between nations are based on economic and social inequalities.

The inequalities of dependent capitalism find expression in two urban dimensions: inter-metropolitan inequalities (between urban and rural areas, and between metropolises and smaller cities and towns); and intra-metropolitan inequalities (between rich and poor areas within the metros). Growth of the dependent metropolis is spurred by the process of accumulation of capital, labor and poverty, but represents an historically progressive development. The standard of living in the metropolis is higher than in rural areas, towns and cities. The main factor in metropolitan growth is natural increase of the population, but migration, political and military factors play a crucial role in certain periods and circumstances. In the following we shall explore these propositions further.

The dependent economy revolves around the export of commodities *and capital* (commodity export by itself does not necessarily bring about dependency, if exchanges are based on equality). The productive forces (especially labor, technology and infrastructure) are relatively undeveloped, even though the most modern technology is readily available and may be used in limited enclaves and industries. The level of subsistence of the working-class population is substantially below the level in the developed nations, though there is wide (and growing) variation among developing nations. The class structure is relatively less developed, as are the two main classes, the working class and bourgeoisie. Because of this, contending classes seek to control state-owned property, which plays a pivotal economic and political role in development. The nascent bourgeoisie and various elite allies often use the state as a direct means of accumulating profit.

The urban land market often plays a disproportionately large role in the economies of dependent capitalist countries. Urban land is one of the few outlets for profitable investment that is open exclusively to local elites and compradors. International finance capital usually finds little reason to circulate its surplus in urban land abroad when profitable investment may require the kind of knowledge of local conditions they have little interest or incentive to acquire. In the metropolises of some of the most developed Third World countries, such as Buenos Aires, Caracas and São Paulo, active real estate markets have produced skylines and sprawl that mirror the pattern of Northern metropolises.

Dependent capitalism is not a distorted or underdeveloped capitalism. It is the particular form taken by capitalism in the former colonies and very much bound up with capitalist development in North America, Europe and Japan. Without dependent capitalism there could be no developed capitalism. (For classical statements on dependency see Cardoso, 1972; Amin, 1976.)

The dependent metropolis evolved from the colonial city, but with the culmination of the revolutions against colonialism the colonial cities experienced a revolutionary change. Colonial cities were administrative centers and way stations for exporting commodities to the colonizing nations. Their internal structures often reflected the social and cultural forms of the colonizing nations. The colonial powers often founded their own cities, but in many cases they occupied indigenous settlements and transformed them into cities. Sometimes the level of urbanization before colonialism was substantial. To varying degrees, however, the colonial powers destroyed the existing settlement systems by imposing the logic of export for profit on the economies of the colonies (see Gilbert and Gugler, 1984: 11–26). Colonial cities, and dependent metropolises, are by no means homogeneous and the differences among them are often substantial (see Lowder, 1986).

After the revolutions against colonialism, the colonial cities exploded in size and territory, and developed complex internal structures, including suburbs with differentiated social strata. Many nations gradually developed economies based less on primitive accumulation and the export of raw materials and more on the production of commodities for the international market. However, the dependency of the cities on the former colonial countries did not change fundamentally; in this there was a line of continuity with colonialism. The new metropolitan areas also became the main symbols of national independence and central to the consolidation of nationhood. Today they are the centers of evolving national economies, national cultures and national social structures.

Latin American colonies were the first to be liberated, and the first to experience the metropolitan explosion. By the mid-nineteenth century most of Latin America was free from Spanish and Portuguese rule. Even though the United States consolidated its economic and political hold on the hemisphere by the beginning of the twentieth century, Latin American nations attained a measure of political and economic independence that most African and Asian nations did not achieve until much later. Latin America's metropolitan areas grew as centers of national capital whose growth has always been dependent on U.S. tutelage.

Latin America has a far more developed metropolitan structure than Asia and Africa. Asia's colonial revolution was not consolidated until the mid-twentieth century (the independence of India in 1947 was key). Africa's independence did not near consolidation until the liberation of Angola in 1976.

The dependent metropolis is an underdeveloped metropolis only in the sense that modern technology and commodity production for the market are not as pervasive as in the metropolitan areas of North America, Europe and Japan. The internal division of labor, and division of urban functions, are not as complex. The urban infra-structure, consumption, local government and the general quality of life is not as developed, and environmental quality is extremely poor.

In sum, the dependent metropolis is an integral part of the post-colonial international division of labor. It exports value created by its labor, and is the gateway through which national wealth is appropriated by foreign capital.

It is no accident that the metropolitan areas of Latin America are mostly located along or close to the coasts where they facilitate the extraction of natural resources and commodities. This pattern, first established by the colonial powers, was not fundamentally altered by independence, or by the brief period of local industrial development and import substitution from the 1930s to the 1950s; rather, new industries tended to reinforce the export-oriented cities. The exceptions prove the rule: Mexico City is the center of the only Latin American nation bordering on the U.S. via land; Bogotá was for a long time the center of one of the most balanced Latin American regional economies; Asunción and La Paz are the capitals of land-locked nations and relatively small.

Unequal development and urban inequalities

Urban inequalities are bound up with and inseparable from economic inequalities. As stated by Armstrong and McGee (1985: 17):

> The role of cities in both capital accumulation and the generation of dependence, structural inequality and poverty is part of the larger history of the unequal relations existing within and between societies.

Violich (1987) outlines four dimensions of urban inequality in Latin America: social and economic dualism; regional polarization; spatial and function inequalities in the urban structure; and centralization and political favoritism in public institutions. All of these elements are found to one degree or another in Africa and Asia as well as in Latin America. Using the first three dimensions outlined by Violich, let us briefly focus on the main features of urban inequalities (the fourth dimension, political centralization, is dealt with later on in the chapter).

Social and economic dualism. Unequal development reflects the unequal exchange between rich and poor nations, unequal distribution of the surplus from production, and unequal ownership of property. It reflects the chasm dividing the standard of living in developed and developing nations, which allows transnational corporations to pay lower wages and buy commodities for less from developing nations. The application of modern technology, the productivity level, and thus the 'rate of exploitation' (where exploitation refers strictly to the rate of extraction of surplus value, not to oppressive conditions) tend to be higher in developed nations, the reason that most capital is reinvested within the orbit of the United States, Europe and Japan. However, along with greater productivity the working class in developed nations has exacted more benefits, which take the form of higher wages and a higher average level of subsistence. The urban infrastructure, including physical and social comforts, is a relatively fixed part of this subsistence level. The increasing improvement of the urban environment in developed nations, contrasted to the degradation of the environment in developing nations, is a product of unequal economic development.

Regional polarization. The enormous gap between urban and rural conditions is partly reflected in the 'primate city' phenomenon, whereby the metropolis dwarfs other settlements within a nation. For example, Lima, Peru may be considered a primate city because it is ten times the size of Arequipa, the second largest urban settlement. This phenomenon is especially pronounced in former colonial nations where the colonial power concentrated disproportionate resources in building capital cities economically oriented towards exporting commodities for profit. In other words, urban–rural differences are partially – if not largely – reflected in the differences between the metropolis and the smaller cities and towns.

The dependent metropolis tends to be a primate city, but is not always a primate city. In fact, as has been amply demonstrated elsewhere, urban primacy, and the whole question of settlement size, may have little to do with poverty and inequality. (See Gilbert and Gugler, 1984: 30, where they show that many developed countries have primate cities.)

Size differences are not useful indicators of imbalances in metropolitan development. Differences in income, standard of living and quality of life are the main indicators. Gilbert and Gugler summarize evidence for what must be obvious to every observer from developing nations: rural areas in developing countries are more impoverished than urban areas, and 'regional disparities in less developed nations are far wider than those in developed countries' (Gilbert and Gugler, 1984: 28). Lipton also states that over 65 per cent of the poor in the world rely on agriculture for their livelihood (Lipton, 1977: 16).

Inequalities in urban structure. The fundamental inequality in the dependent metropolis is between the overwhelming majority of the population who live in poverty and the minority who have a decent stable living environment. The majority of the metropolitan population in developing nations lives far below the average standard of living of the developed nations, many of them in massive spontaneous communities and shantytowns (Dwyer, 1979). Because of the low level of subsistence in the dependent metropolis, capital is able to keep wages low, both in the dependent metropolis and in the developed metropolis. The Malthusian Iron Law of Wages has gone international (see Chapter 1).

Mass poverty in the Third World includes the many communities often described as 'marginal,' outside the realm of the 'modern' capitalist market. It is true that a large portion of the dependent metropolis survives by producing goods and services outside of 'formal' industrial production. However, the so-called informal sector is nothing more than the poorest stratum of the working class, and in relative terms the people in this stratum usually enjoy a higher standard of living than rural workers. As in many developed metropolitan areas, they work in poorly paid services that make up an essential part of the metropolitan economy; they may be unregulated by government, but are not at all 'marginal' or insignificant (see Perlman, 1976; Bromley and Gerry, 1979; Portes, 1985; Portes, Castells and Benton, 1989). This is not to say that this sector can in any way be strategic to economic development, as suggested by some economists who glorify unfettered small-scale entrepreneurship to justify government deregulation of business.

Another key aspect of urban inequality is displacement. As soon as labor migrates into metropolitan territory it is threatened by displacement. As in the U.S. model, the private real estate market bids up the cost of land in central locations, and forces the expulsion of low-income workers in favor of office buildings, stores and new housing for the professional and technocratic strata. Struggles over displacement parallel and intensify class and income divisions because the dependent elites and bureaucratic and technical strata are heavily invested in real estate (productive investments are tied up by international capital).

Displacement in the dependent metropolis differs from the U.S. model in that it mostly affects suburbanites rather than central city dwellers. Immigrants from the countryside often settle in shantytowns and squatter settlements located in the urban periphery beyond the existing network of urban services. If the land they happen to occupy has any significant real estate value, they become the target of eviction. They are more likely to be tolerated when the land has little exchange value – for example, on the banks of rivers subject to flooding (as in Lima), on steep slopes subject to volcanic activity and landslides (as in San Salvador), and near noxious public facilities like waste-treatment plants.

Shantytowns are just one step from homelessness and the dividing line between the two is often not clear; a shack made of cardboard with no water, electricity or plumbing is little more protection than street sleepers have. There are perhaps millions of homeless people around the world, and many more living in shantytowns (no precise accounting exists). In Calcutta, there are an estimated 300,000 pavement dwellers, many of whom are refugees from Bangladesh with little possibility of securing permanent housing, and who stay in the same place for years with nothing but sheets of plastic to protect them from the elements. The pavement dwellers are the poorest housed sector in a metropolis with millions of poorly housed people; many of them are employed and are not significantly more economically marginal than shantytown dwellers.

Entire neighborhoods are often displaced in developing nations to protect elite property rights, facilitate monumental public works, and maintain political control over the poor. The uprooting of spontaneous communities and shanytowns also dampens demands for greater government expenditures for urban infrastructure, expenditures that create further demands for wage increases. Among the most dramatic examples of displacement are the forced relocations of squatter settlements in Mexico, Peru, Brazil and other Latin American countries, where they have given rise to large popular protest movements (see Castells, 1983: 190–209). The government-sponsored 'urban renewal' of Lagos (Marris, 1960) became one of the most notorious cases, but there have been many less publicized examples. Dwyer (1979: 79–85) presents the case of the clearance of thousands of squatters from the Intramuros area in central Manila to facilitate development of a national cultural and civic center.

Accumulation in the dependent metropolis

The metropolitan structure in dependent countries is fundamentally shaped by the international division of labor. The production of commodities is now spread across the planet, and the most tedious and lowest-paid steps in the production process are located in the dependent metropolis. Taking advantage of differences in labor costs and low transport costs, transnational corporations can now manage an assembly line spanning the globe (see Taylor and Thrift, 1986).

Since the 1970s, capitalist production has begun to take on more pronounced aspects of a global assembly line. Many have associated this with a 'new international division of labor' (Palloix, 1977; Frobel, Henrichs and Kreye, 1980; Henderson and Castells, 1987; King, 1990). This phenomenon has been triggered by deindustrialization in the developed countries – the closing of smokestack industries and certain labor-intensive operations – and the flight of some industries to developing nations. It began with the growth of sweatshops and sweatshop economies in, for example, Taiwan, South Korea and the Philippines. But in its new phase it entails the export of labor-intensive assembly operations within the new high-tech electronics and communications industries. At the same time, the dynamic metropolitan areas in the developed countries serve less as centers of production and more as global centers for international capital.

Distribution and consumption are also to some extent becoming internationalized, as transnational corporations have opened up new markets in the metropolitan areas of developing countries. Armstrong and McGee make the link between accumulation and consumption in a forceful way:

> Cities are, stated simply, the crucial elements in accumulation at all levels, regional, national and international, providing both the institutional framework and the *locus operandi* for transnationals, local oligopoly capital and the modernizing national state. In offering advantageous conditions for capital, cities, particularly the large metropolitan areas, act as the central places for a process leading to an increasing concentration of financial, commercial and industrial power and decision making. On the other hand, cities also play the role of diffusers of the lifestyles, customs, tastes, fashions and consumer habits of modern industrial society. These two processes of centralization and diffusion are, of course, not contradictory

> for the expansion of markets is a necessary part of growth in the capitalist system.
>
> (Armstrong and McGee, 1985: 41)

Despite the growing globalization of production and consumption, it has yet to be demonstrated that the new international division of labor is *qualitatively* different from the division of labor that has prevailed everywhere since the end of colonialism, in which national economic and political institutions throughout the world continue to play an important role. The analysis here diverges from the world systems approach (see Chase-Dunn, 1984; Timberlake, 1985), which leads to an underestimation of the durability of national systems and overestimation of the significance of the global assembly line phenomenon.

Capital and industry exported from the developed nations is still going primarily to other developed nations. Japan and Europe are investing heavily in the United States; their investments in developing nations, though significant and increasing as a proportion of the total, have not altered the post-war economic order in which capital, manufacturing and consumption of the surplus is concentrated in the developed capitalist countries. Thus, for example, while manufacturing, particularly the assembly stage, in the leading industries of auto and electronics has now spread to countries like Brazil, the vast majority of value added to world commodities still originates in the developed nations. While new enclaves of capital accumulation have emerged in the Third World, such as the Republic of Korea, Taiwan and Hong Kong, they still do not account for more than a small proportion of total capital investment. Saskia Sassen, while showing how the rate of direct foreign investment by developed countries in developing countries has grown, particularly since the 1970s, acknowledges that 'most of [the direct foreign investment by the major industrial countries] was in developed countries' (Sassen, 1988: 99).

Since the 1970s, when U.S. monopoly control over the accumulation process began to dissipate, capital has increasingly accumulated in Japan and the European community. These two economic powers have also received an increased share of the capital flowing from the dependent nations. Indeed, the new wave of growth is taking the form of European and Japanese expansion into territory previously monopolized by the U.S., including Latin America, Uncle Sam's 'back yard.'

The effect of these economic changes on the dependent metropolis remains to be seen. The tendency may well be toward slower metropolitan growth given the nature of the accumulation model in Western Europe. This model is less akin to primitive accumulation,[1] which exploits large amounts of new labor within a very short period of time and therefore encourages rapid rural–urban migration and urbanization. European capital exhausted this model in the colonial era. However, the Japanese accumulation process appears to be more oriented towards primitive accumulation (even within Japan, where new industrial centers at the periphery of metropolitan areas are filled with sweatshops). In coming years, the impact of Japanese economic expansion on urban growth may be dramatic as Japan's orbit of reinvestment expands.

As world capitalism is becoming increasingly polycentric, new centers of accumulation have arisen in the developing world. These centers tend to be among the most highly urbanized, and their metropolitan areas the most modern in appearance and structure. The Gulf States, Saudi Arabia, Israel, Korea, Hong Kong, Taiwan, Singapore, Argentina, Brazil, Mexico and Venezuela are now major centers of capital accumulation. In addition there are many tiny tax havens specializing in conspicuous consumption, such as the Bahamas and Cayman Islands, though they are generally outside the spatial limits of the metropolis.

Some of these new centers, particularly the Asian centers, are enclaves for primitive accumulation in production, intensive real estate investment, and regional banking. The Gulf States are enclaves built around the extraction of oil and the distribution of petroleum profits among a limited group of clans. As with some of the Asian enclaves, their growth depends on the exploitation of a small group of immigrant labor living far below the level of opulent consumption that characterizes the ruling elites. The enclave character of these centers is highlighted by the fact that the vast majority of the population in their regions lives outside these nations. The wealth and higher standard of living they are noted for have not altered the mass character of poverty in their regions.

The indebtedness of the four largest Latin American economies, particularly Brazil, to the older centers of capital speaks to the limitations on the independent development of any large Third World nations. The extent to which these nations can command reinvestment of the surplus from export production and develop their independent, internal accumulation processes depends on the emergence of a new international economic order. In the existing order, the surplus is drained to the international centers of capital, which now regulate the flow through debt financing.

A truly national process, if such a thing is possible in an economy closely linked with, and indebted to, transnational capital, might possibly produce a more decentralized, balanced settlement system with larger interior cities. However, the direction appears to be toward greater dependency, not toward a maturation of independent national economies. In many countries, the latter possibility is being foreclosed by the debt crisis. Creditors have forced governments in even the more developed countries to initiate privatization policies, which often mean the transfer of national assets to transnationals. Governments have had to make investments that reinforce the existing division of labor and dependent urban structure.

Metropolitan growth: economic versus political factors

While in general capital accumulation produces urbanization, this is not a sufficient explanation for urbanization in specific instances, particularly the rate and structure of urbanization within individual nations and regions. For this, two factors need to be considered: the rate of natural increase (births less deaths) and migration. As noted in Chapter 1, these factors are largely determined by economics, but not always. Often, political and natural events shape changes in natural increase and migration.

The rapid growth of metropolitan areas in developing nations is due mostly to the rate of natural increase, which has gone up with a decline in mortality rates. Samuel Preston aptly summarizes the findings of demographers in this respect:

> 1. The rate of change in the urban proportion in developing countries is not exceptionally rapid by historical standards; rather it is the growth-rates of urban populations that represent an unprecedented phenomenon.
> 2. Urban growth through most of the developing world results primarily from the natural increase of urban population.
> 3. Among the factors that influence the growth rate of individual cities, national rates of population growth stand out as dominant in inter-city comparisons.
> 4. Urban growth in developing countries has typically not been associated with a deterioration in industry/urban ratios.
>
> (Preston, 1988)

While general population trends can explain the rate of urban population increase, migration and political/military factors can better explain why some individual settlements, particularly metropolises, grow proportionately faster than the average. Metropolitan growth attributable to migration is significant, but migration usually occurs within limited periods of time. Since most demographers use very low thresholds to define urban (e.g., settlements of 20,000), general conclusions such as Preston's above obscure the qualitatively distinct situation with respect to metropolitan areas. In the following we will consider how the metropolis tends to grow due to migration and political/military factors.

Migration to the dependent metropolis is spurred by 'push' and 'pull' factors that reflect regional economic and social inequalities. Pushing people off farms are, for example, poverty, unemployment, and lack of services and cultural opportunities. Pulling people to the metropolis are jobs, housing, health care, educational opportunities and broader cultural opportunities; despite urban poverty, conditions tend to be better in the metropolis than the village (see Gugler, 1988: 74–92). Not all push-and-pull factors are equal. The central factor is economic (see Gilbert and Gugler, 1984: 52). More precisely, the central factor is capital's offer to employ labor – a 'pull' factor.

Along with the accumulation of working population, there is an accumulation of a labor reserve. The new factory not only replaces small-scale commodity production with wage employment, but leaves large sections of the population unemployed. The *barriadas*, *callampas* and *bidonvilles* are ready reserves for international capital to utilize in its global factory, whether it is located in the developing nation or the developed nation. Sassen (1988) has shown how capitalist expansion in developing countries tends to encourage migration to developed countries. The huge labor surplus in the dependent metropolis is as necessary to capital as the small proportion of the population actually employed in commodity production for exchange on the international market. For their low level of subsistence suppresses wages and the expenditures that capital must make for the metropolitan infrastructure.

The owners of capital allow the dependent metropolis to grow spontaneously, without public investments that would be required in the metropolis of the developed world. Water and sewer systems, street surfacing, electricity, parks and recreation, schools and health facilities are not financed by the global factory's surplus, which is exported, but instead by the meager accumulations of local capital, which are unable to support a modern infrastructure. The dependent metropolis does not have the physical infrastructure necessary for a better quality of life, nor the economic means to guarantee it, nor the institutional infrastructure – planning and regulatory mechanisms – to enforce environmental standards.

Consequently, the micro-environment in the dependent metropolis is the most dangerous and polluted in the world. For example, carbon monoxide levels in Mexico City far exceed the averages in North American metropolises. Mexico does not have the surplus capital that would be needed to replace the private automobile with mass transit, and is locked into a mode of accumulation that forces it to rely on environmentally damaging petroleum energy. Mexico imports its model for metropolitan development from the North, but cannot afford to import the means to control it. Although the rate of vehicle ownership is lower than in the North, cars are older and have highly inadequate emission systems, and fuel is cruder than in the North.

The Bhopal disaster is one of many environmental disasters in developing countries that can be traced to the relative freedom corporations enjoy in locating noxious production facilities in a way that allows low-wage residential areas to spring up nearby. For example, Brazil's main petrochemical center, Cubatao, is located less than 80 kilometers from São Paulo. Every day about 1,000 tons of pollutants, including benzene, carbon monoxide, ammoniac and other toxics, are spewed into the air. According to Finquelievich (1990), 'in the nearby "favelas," infant mortality reaches 35 per cent' and '80 per cent of the Cubatao children suffer from respiratory problems.' Another example provided by Finquelievich is the nuclear generating facility located in Ezeiza, a low-income area near Buenos Aires with a high water table susceptible to contamination by radiation.

Environmental degradation is not only associated with metropolitan development. There are many examples of vast devastation of natural resources by rural development, including agriculture – the deforestation of the Andes, desertification in northern Africa, and the destruction of the Amazon rain forest, to name a few. In the latter case, the migration of settlers from urban to rural areas, with government assistance or acquiescence, has helped create one of the greatest environmental disasters in the Americas. Settlers typically strip the forest, farm intensively until the natural nutrients in the soil are used up, and then abandon the land. In the Amazon gold-mining towns, two pounds of toxic mercury runs off into streams for each pound of gold extracted. In addition, workers in the gold mines are not protected from the inhalation of mercury vapors.

The issue of environmental protection is qualitatively distinct in developed and developing nations. This is one of the points often overlooked in discussions of 'sustainable development' proposals which aim at environmentally sound development and consumption strategies in the Third World (see Cadman and Payne, 1990). To be sure, elimination of pollution in both the developed and developing nations is an urgent and humane objective. The developed nations, however, are the main contributors to world pollution, and they need greater energy efficiency most urgently, while developing nations suffer some of the worst consequences of global environmental contamination. Insofar as measures to achieve greater energy efficiency in the developed nations cut down on the extraction of raw materials from developing nations, they can prevent their depletion. However, from the point of view of the developing nations, unless inequalities between the two worlds are also eliminated, pollution control in the developed world will improve resource efficiency and the quality of life there without necessarily improving life in developing nations. In fact, more resource-efficient production and consumption in developed nations can cause great economic damage by lowering world market prices of the few goods poor countries have to sell. And there is also evidence in the developed nations of a trend toward discarding the most noxious waste materials, which cannot be economically recycled or disposed of within their own borders, in Africa, Asia and Latin America.

Migration to the metropolis is not always a direct result of economic 'push' and 'pull' factors. There are many examples in history of political and military policies that 'push' and 'pull,' sometimes in the same direction as spontaneous economic trends and sometimes in opposite directions. The role of the state may at times be decisive. There are also many examples of natural disasters and epidemics which affect urban fertility and mortality rates or force migrations.

Israel is a case in which political/military factors are decisive. The dependent population in Israel – some two million Palestinians, Israel's low-wage labor reserve – is concentrated in two major semi-urbanized areas, the West Bank and Gaza, separate from the majority of the Israeli population. When the State of Israel was founded in

1948, its rulers changed the Zionist policy of promoting dispersed rural settlements to one of promoting urban settlement (Troen, 1988). They promoted the immigration and settlement of millions of Europeans in the coastal conurbation made up of the cities of Tel Aviv, Haifa and Jerusalem, and numerous medium-sized cities. This was a military policy that determined urban policy. To facilitate new settlements, millions of Palestinians were expelled from rural areas to the West Bank and Gaza, and other Middle Eastern nations (almost two million live in neighboring Jordan). Palestinian farmland was taken over by Israeli settlers and farmers. By depriving them of land and citizenship, and dispersing them, Israel expected to eliminate an obstacle to the development of its metropolis. Thus, the metropolis in Palestine is the product of forced migration.

Another case in point is Vietnam. During the Vietnam War, the United States carried out a policy of forced urbanization and 'rural pacification.' This was primarily a military policy aimed at forcing the rural population, largely loyal to the liberation movement, into U.S.-controlled territory. At first, the U.S. tried establishing 'strategic hamlets' under their control. Then they undertook saturation bombing and a scorched-earth policy in the rural areas, which forced a massive migration to Saigon (now Ho Chi Minh City). The U.S. created a bloated primate city. At the same time, saturation bombing in North Vietnam forced the dispersal of population from the capital city of Hanoi. After the war, a united Vietnam inherited a settlement structure with a dramatic imbalance between North and South, rural impoverishment and the bloated metropolis of Saigon (see Forbes and Thrift, 1987: 98–128).

Another example is Bangladesh. Within the five to six years following partition and establishment of Bangladesh, three and a half million people migrated from this mostly rural area to the Indian state of West Bengal. Most of them settled in the Calcutta metropolitan area, where they and their families now make up about one-third of the population. Most of the migrants were Hindus and other ethnic minorities who fled the mostly Muslim state of Bangladesh.

Finally, we can look at Nicaragua during the Contra war. Between 1980 and 1987, Nicaragua went from 53.4 per cent urban to 63.6 per cent urban. The average annual urban growth rate was nearly twice the national growth rate (Higgins, 1990; the author uses a different definition of urban than we have followed in this text). The rapid migration was due principally to the war in the countryside and the economic crisis caused by the United States blockade.

These cases dramatize how metropolitan development can be caused by more than just the spontaneous operation of 'push' and 'pull' factors in the markets for capital and labor. The dependent metropolis results from economic factors and may serve the interests of capital, but political factors may also play a crucial role in shaping the metropolis, both with respect to its internal structure and its role in the national and international systems of settlements.

UNEQUAL METROPOLITAN DEVELOPMENT AND EQUALIZATION POLICIES

Metropolitan planning has a particular significance for both developing capitalist and socialist nations. It can be used either to assert and consolidate an emerging national identity or rationalize a dependent international status. It can be part of the struggle to overcome the legacy of inequality, or it can be an acknowledgment of the supremacy of the profit principle over social justice. The amelioration of urban and regional inequalities can help forge a national identity by reducing the contradictions between urban and rural areas. To accept inequalities as inevitable, or as only susceptible to minor adjustments, is to succumb to the logic of the unequal international economic order.

In concrete terms, improvement of the urban infrastructure in the dependent metropolis means raising the standard of living that international capital must base its investment and wage strategies on. By maximizing the returns to the nation's labor and capital, developing nations can come closer to the standard of living of the developed nations. This, at least, is the hope, and the rationale behind the more aggressive and forward-looking planning efforts in developing nations that transcend the immediate economic possibilities of the urban marketplace.

This contradicts the approach of analysts such as Michael Lipton (1977) and Alain deJanvry (1981), who see rural development as strategic. Lipton, arguing against what he calls the 'urban bias' in development strategies, states:

> Only on the basis of a tolerable level of living for a mass agriculture of small farmers can most poor countries construct, speedily and efficiently, a modern industrial society.
>
> (Lipton, 1977: 23)

Lipton proposes greater investments in rural development and concrete measures to bring government closer to the rural masses. A structuralist who sees little benefit in linking urban and rural development, deJanvry calls for rural self-sufficiency. As shown in the rest of this chapter, this anti-urban approach, in practice, has not produced the desired rise in living standards.

Metropolitan planning in Africa and most of Asia is especially concerned with relations between city and countryside, and relations between metropolis, city and town. However, metropolitan planning in Latin America, which is more urbanized, tends to be focused on planning within the metropolis.

Metropolitan planning by itself is politically neutral. It can be revolutionary or conservative. It can seek to strengthen national independence or serve neo-colonialism and imperialism. It can seek to establish national values and priorities or follow the dictates of 'rational planning' from Europe and North America.

The most problematic schemes for eliminating inequalities narrowly focus on the need to decentralize the urban structure. Decentralizing a highly centralized urban and economic structure has long been seen as a means of promoting national development as well as eliminating inequalities. It is supposed to stimulate an internal market that can be relatively independent of external capital through import substitution. It is supposed to eliminate differences between city and countryside. Decentralization is supposed to be a means of reducing rural dependency, rectifying uneven regional development, and eliminating urban primacy.

However, decentralization schemes are often misdirected, both in theory and practice, because they do not pose the problem as essentially one of *economic inequality* within the context of uneven development. Rather, decentralization is usually considered an essentially urban, or spatial, problem. The problem of overcentralization is thereby measured by divergence from some ideal notion of even development between city and countryside, large and small city, metropolis and rural area. In this section, I will attempt to show that development strategies are best undertaken as strategies for *the reduction of economic inequalities* rather than *urban decentralization* strategies.

Central economic planning within the context of independent national development policies provides the best opportunity for a more equitable urban planning practice. A greater degree of control over the nation's human and natural resources is a precondition for directing and influencing their location and use. Regulation of external factors is the first step in regulating internal factors.

The ability to plan and regulate the dependent metropolis, however, is not strictly a matter of political will. Experience indicates that central planning by itself is not enough. Planning is severely constrained by external factors and by objective limits on resources. The truth of these statements may seem obvious, but it is a truth most easily forgotten by planners. Failure to sufficiently take into account objective constraints opens the door to schemes that promise but never deliver regional equality, and can ultimately expand inequalities. This problem is especially present in developing socialist countries with central planning institutions. Because of the inexorable role of the marketplace, both national and international, most avowedly socialist countries have mixed economies with varying degrees of actual control over national development. Those that chose to completely ignore markets are now in the most serious of difficulties.

In the following sections, let us examine several common metropolitan planning strategies that involve some form of decentralization. These are: (1) creation of growth poles around new centers of industrial production; (2) establishment of policies to balance regional development; (3) decentralization of administrative functions by creating a new capital city; (4) strengthening of local government; (5) promotion of local self-help initiatives; and (6) centrally planned equalization. We will try to show that the centrally planned equalization strategy holds the greatest promise of success, but only when the objective laws of urban and economic development are fully taken into account.

Growth poles

The notion of promoting equitable regional development by establishing industrial growth poles in less developed rural areas has a long history and has taken many forms. In the 1960s, many international aid institutions toyed with these ideas, but very little was actually done. In part, these agencies responded to the somewhat successful establishment of new towns in the planned socialist economies. For example, in the post-war period the Soviet Union built about a thousand new towns around industries, both in the European part of the country and in the Far East, where new town development was an integral part of opening up regions to new development. Very few of these new towns became metropolitan areas, but all provided industrial jobs outside the orbit of existing metropolises (the most notable one that became a metropolis was Togliattigrad in Russia, built around a giant auto factory, and now well over 500,000 population).

The earliest proposals for growth poles in the West simply called for establishing industries in rural areas. An early example was Ciudad Guayana in Venezuela (see Rodwin, 1969). John Friedmann (1966) and Antoni Kuklinski (1972) came up with more complex proposals for establishing a 'growth center policy.' This policy would place a few dynamic industries in less developed cities or towns to spur more equal regional development across the national landscape.

Many, including Friedmann, soon realized the problems with this conception (see Friedmann and Weaver, 1979). First of all, the concept is so broad it can cover a wide range of new development experiences, from building new towns from scratch to dispersing large industries to small cities. But the growth center idea fell out of fashion with the international aid establishment because it required a level of central economic planning unacceptable in the capitalist world. The international aid establishment, which dominates most bilateral and multilateral aid programs, has consistently demonstrated that it is not prepared to subsidize national planning efforts that interfere with the agenda of transnational corporations.

At the theoretical level, growth pole strategies never adequately dealt with the problem of urban economies of scale. Existing larger settlements probably offer more efficient use of economic resources (see Gilbert and Gugler, 1984: 177–8). The metropolis in particular provides many scale economies in the production sphere, and in social and physical infrastructures. There may also be diseconomies of scale as settlements reach a size whereby the economic and social costs of building and operating the infrastructure exceed benefits. However, the debates over scale economies are inconclusive, and indicate that diseconomies may exist in some services but not all. Debates over an 'optimal city size' at which scale economies are optimized are also unresolved.

One of the earliest experiences with growth poles in the capitalist world was in Italy. Since the founding of the republic in 1861, Italy has been sharply divided between an industrially developed North and underdeveloped South, the *Mezzogiorno*. The South has been less urbanized, dependent on the industrial North for goods, and politically atomized. Incomes are lower in the South, which until only recently has been the North's main labor reserve (in the 1980s, Italy began to import labor from northern Africa).

Antonio Gramsci first identified 'the Southern Question' – unequal development between North and South – as the key to consolidating the Italian republic (Gramsci, 1957). As one of the last European nations to unify, and one of the poorest, Italy's future would depend on a dramatic, revolutionary resolution of the problem. The Fascist government between the world wars maintained industrial concentration in the North, undertaking some limited land reclamation projects in rural areas and urban renewal projects in urban areas, neither of which fundamentally restructured the urban system. After World War II, the Christian Democratic government initiated the 'Cassa del Mezzogiorno,' a program that included the establishment of growth poles in the rural South.

Major new industries located in the South include petroleum refining, chemicals, shipbuilding and auto assembly. There have also been many smaller labor-intensive industries such as textiles and electronics assembly. One of the effects of this growth has been to encourage urbanization of the South, but not necessarily around the new industries. Most of the new industry has instead added to the growth of existing metropolitan areas. Industries far outside metropolitan areas have become 'cathedrals in the desert,' yielding few jobs, with practically no multiplier effect on the local economy. Furthermore, capital investment in the South, even at its height, never came near investment in the North, so that emigration of southern labor to the North far outpaced migration to the southern growth poles (*Critica Marxista*, 1989).

Another example of growth pole planning is South Africa. While the major institutions of apartheid in South Africa are now crumbling, many of its features continue to exist, particularly racial segregation. After 1948, South Africa carried out a national policy of 'growth points' and 'influx control.' The growth point strategy promoted development around industrial growth poles (Geyer, 1989). This went hand in hand with the establishment of bantustans, segregated African communities serving as suburbs for the growth points (although many at large commuting distances). It was also accompanied by the establishment of segregated townships on the outskirts of the once-white metropolises of Cape Town, Durban, Johannesburg and Pretoria. Migration from the bantustans to these townships was limited by 'influx control.' The Urban Areas Act of 1923 and the

Group Areas Act of 1950 codified the segregated system (see Morris, 1981).

The strategy of apartheid was to keep most of the African population on the bantustans and limit migration to the metropolis through 'influx control.' By requiring every African to carry a passport and obtain government approval for migration to the metropolis, the unequal system of remunerating black and white labor was maintained. The lowest paid labor reserve, including women and children, remained in the bantustans, far from the metropolis, and the slightly higher paid and more active labor reserve was allowed to live in townships. In South Africa in 1990, black wages were 21 per cent of white wages in construction, 30 per cent in manufacturing and 46 per cent in the finance sector (*Newsweek*, 1990). Segregation of blacks from whites guaranteed that black workers would not receive the same level of benefits from the urban infrastructure as white workers did, thereby further widening the differences between the two and heightening the level of exploitation of black labor. It is difficult to imagine how apartheid's separate wage structure can be maintained if the black and white communities are mixed and the same standard of urban living were to apply to all races.

In order to enforce this system, the South African government frequently removed Africans who migrated to the metropolis without official authorization. In the last twenty-five years at least three and a half million people were uprooted from the townships by force, and thousands died for resisting. The black settlement of Crossroads in Cape Town was repeatedly uprooted, only to have residents return and rebuild. Other settlements, like District 6 in Cape Town, were completely demolished, supposedly to make way for a new white neighborhood.

By the late 1980s, it became clear that the 'growth point' strategy and 'influx control' had failed on their own terms. Migration to the metropolis was not stopped, despite use of the most repressive methods. In the townships, the average occupancy rate of the shacks was ten persons. In the Transvaal region, the government acknowledged 200,000 illegal shacks, 30,000 of them on white-owned land (Black Sash, 1989: 8–9). As a consequence, the metropolitan areas were mostly black, and there was substantial movement to and from the metros and the bantustans.

As early as the mid-1980s, the South African government began to acknowledge the fact that its strategy had failed. This coincided with a general realization that, due to the rising power of labor, the level of national and international protest, and the South African economic crisis, capital accumulation under apartheid was no longer tenable. South African capital began to prepare for the transition to what they believed could be a more 'normal' process of capitalist accumulation without apartheid. In this model, the unequal labor and spatial structures would be maintained by the 'normal' functioning of the real estate market.

In the 1980s, the South African government dropped passport controls; arrests went from 262,904 in 1983 to zero in 1986 (Sutcliffe, Todes and Walker, 1990: 11). They all but gave up on 'influx control.' While some forced removals continued to occur, the government declared its intention to abandon this policy. Repatriations from the townships to the bantustans were virtually ended. The Group Areas Acts were eventually repealed.

South Africa's new policies were laid out in a 1986 government White Paper, which called for 'orderly urbanization,' and some have already been implemented. They include privatization of black housing and services (begun in the early 1980s), 'negotiated resettlement' instead of forced removal, and the devolution of responsibility to local authorities faithful to the principles of racial separation.

One indicator of the effect of these policies is the new wave of rent strikes in the townships, which began in Sebokeng and Sharpeville in 1984, and Soweto in 1986, and has come to involve hundreds of thousands of households. The strikes responded to government attempts to make blacks pay for housing and services through increased rents, a tactic undoubtedly intended to pave the way for the sale of all public housing to individuals. There are also the first examples of forced removals by private landowners (Sutcliffe, Todes and Walker, 1990).

Balanced regional development

The point of departure for this strategy is the *central place theory* of urban geography (see Christaller, 1966). This theory posits a 'normal' hierarchy of settlements in space, in which the largest and most important settlements are one of three kinds: marketplaces for goods, transportation nodes or administrative centers. The relationship between the size of central places and lesser settlements is supposed to be quantifiable and, unlike the primate city phenomenon, pro-

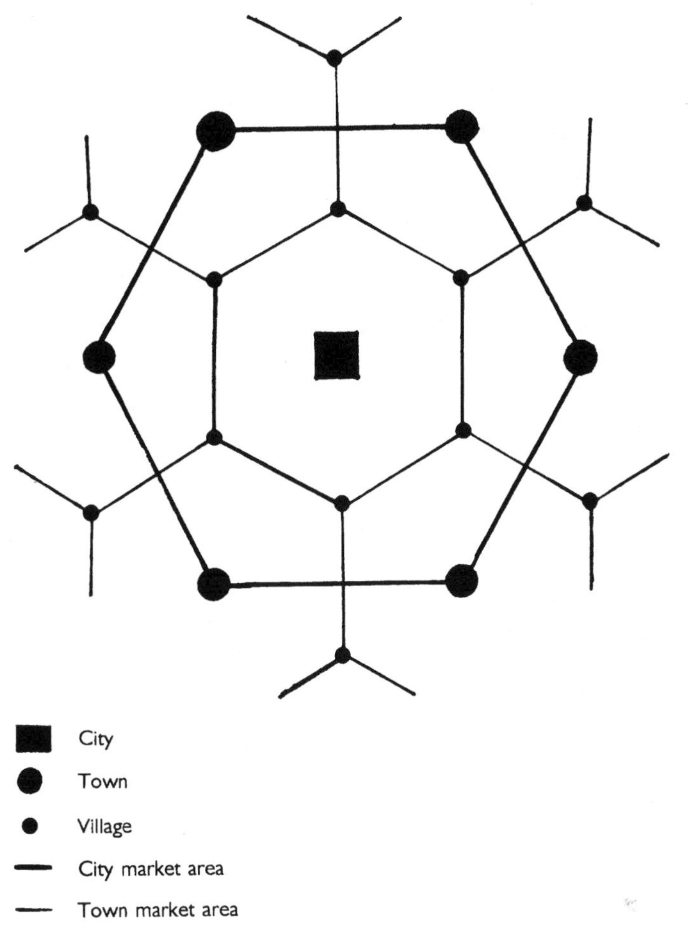

Figure 3.1 Christaller's market principle produces a regional settlement pattern similar to the one above.

portionate. Central places should not be disproportionately large, and there should be a whole range of urban settlements of different sizes. Based on his studies of southern Germany, Christaller's work has been used to demonstrate that where central places are disproportionately large, small and medium-sized settlements need to be built up (see Rondinelli, 1983). The strategy is then to support market, transportation and administrative functions in small and medium-sized cities.

The problem with this approach is that it is rooted in an abstract ideal theory of geography, and equates a particular kind of spatial uniformity with economic equality. It confuses functional and geographical categories. Economic development does not arise from geographical form or 'normal' dimensions. A regular size relationship among settlements does not necessarily mean the economic relations are based on equality.

Christaller's central place theory can be used to describe a particular urban pattern in central Germany, or various other regions around the world, but does not explain much about the main trend of unequal metropolitan development in the twentieth century. There are so many objective economic, political and natural constraints that alter the central place pattern that the ideal construction is meaningless in the practical world. Central place patterns appear in a relatively uninterrupted flat geographical plane, in a relatively diverse and undeveloped rural marketplace, and in an administrative system that enjoys ample resources (see C.A. Smith, 1976, for detailed analyses of central place theory).

To achieve the desired uniform geographical distribution of central place theory, one must fall back on more specific development strategies. Following the market principle, towns and cities can receive assistance in acquiring the infrastructure and technical means to participate in the regional market. To strengthen the transportation principle, roads and transit infrastructure can be built. To strengthen the administrative pattern, new government offices can be put in place. Rondinelli and Ruddle (1978) put forth as a solution the establishment of 'urban functions in rural development' so that all settlements above given threshold sizes have urban services that correspond to settlements of these sizes.

This approach adds up to an aid program for infrastructure very similar to the kind commonly promoted by the international aid establishment. While such programs can improve the quality of life in less urbanized areas, however, insofar as they reinforce the local marketplace, they reproduce all of the inequalities of the marketplace. Some believe that building up the towns and cities amounts to a sort of internal colonialism (Gonzalez Casanova, 1970). Indeed, when the underlying economic relations are unequal, central place hierarchies may be less like regular geometries and more like trees with thin branches that nourish large roots, what Klara Kelley calls 'dendritic systems' (Kelley, 1976). The market's tendency is to centralize resources as it develops, and to expand and link up with the international market. The roads drain rural areas of population and resources and pull them toward the export-oriented metropolis; and undemocratic centralized power is unaffected by a decentralized government bureaucracy. Indeed, creating urban functions in this economic environment only reinforces the power of urban elites.

Table 3.1 Proposed new capital cities in Africa, Asia and Latin America since 1950

Year initiated	*Country*	*New capital*	*Former capital*
1956	Brazil	Brasilia	Rio de Janeiro
1957	Mauritania	Nouakchott	Saint Louis
1959	Pakistan	Islamabad	Karachi
1961	Botswana	Gaberone	Mafeking
1963	Libya	Beida	Tripoli
1965	Malawi	Lilongwe	Zomba
1970	Belize	Belmopan	Belize City
1973	Tanzania	Dodoma	Dar es Salaam
1975	Nigeria	Abuja	Lagos
1983	Ivory Coast	Yamoussoukro	Abidjan
1987	Argentina	Viedma/Carmen	Buenos Aires

Source: Adapted from Gilbert (1989) and other sources.

Capital city development

Another strategy for decentralization is to relocate administrative functions from the metropolis to less developed regions. While some countries such as South Africa, Chile and Bolivia have split administrative functions among urban centers, the most dramatic attempts at administrative decentralization involve relocation of the national capital. The capital city in formerly colonial nations is often a symbol of national independence and future development. Therefore, creating a new administrative center distinct from the colonial capital is important. The prime examples of new capitals in recent decades are in dependent capitalist countries. These include eleven new cities or major town expansion programs proposed since 1950 (see Table 3.1).

Experience with these capitals shows that if the new capital is to become more than just a symbol it must be planned as a metropolis with a fully developed economic life. Some of these would-be capitals, like Beida in Libya and Viedma in Argentina, never got beyond the proposal stage. Others, like Abuja in Nigeria, have stalled due to changes in the international market unfavorable to the nation's export products. Lauchlin Currie notes that in countries where seats of government were moved, such as Australia, Canada, Brazil and the United States, the major cities did not lose their importance and continued to grow (Currie, 1976: 55).

Some new capital cities have contributed to changing regional and

Figure 3.2 A shantytown in Mexico.

national economies. Brasilia and Islamabad, for example, have had a definite impact on internal economic development, even if they developed at a pace and in a way unanticipated by the planners (see Epstein, 1973, on Brasilia). Tanzania's Dodoma is growing gradually, in accordance with the national policy of encouraging decentralized national development through the cooperative *ujaama* movement. When capital city development is more than relocation of an administrative center, and part of a broader national and regional development strategy, it can influence the form of the national settlement structure. Earlier successful new capital cities include Ankara in Turkey, which became the capital in 1923, and New Delhi in India, which was inaugurated in 1931.

Local governance and decentralization

It is often said that non-metropolitan areas do not develop because political power, and with it economic resources, are centralized. The decentralization of political power and strengthening of local government is commonly proposed as a remedy.

Gilbert and Gugler (1984: 170) state that 'Centralization is a much abused term. The essential problem is that it is used interchangeably both in a functional and in a geographical sense.' They distinguish between functional centralization (the concentration of political power) and spatial concentration (the geographical distribution of power, population, economic resources, etc.). Thus, it is quite possible to have decentralized political power but still have concentrations of population, economic resources, services, etc.

A further distinction must be made between political power and administrative institutions. It is possible to have centralized political power and decentralized administrative units, and it is also possible to have decentralized political power and centralized administrative units. In other words, the presence of branch offices in every province does not necessarily mean decentralization of political power. Likewise, a system giving wide political powers to local authorities may at the same time have a highly centralized, national-level administrative structure.

Decentralization may well be desirable and necessary for national development, but it is not the *key* to national development. A sound national development strategy, based on comprehensive economic and social policies that seek to eliminate inequalities, should usually have a decentralization component, but without the overall strategy decentralization will go nowhere.

Although it may seem to be a paradox, centralized planning can provide the best conditions for decentralization in many countries, because to decentralize the means of production and the social surplus it is often necessary that they first be centrally owned. Centrally planned equalization policies can help balance uneven development and eliminate inequalities.

Over sixty years ago, Jose Carlos Mariátegui (1928) elaborated the principle of 'democratic centralism' as the basis for a genuine decentralization and social equality. Mariátegui criticized half-hearted reforms in Peru that only decentralized government offices. He believed that in a nation with a wide gap between city and countryside decentralization should be based on a truly national government representing the interests of both the working class and peasantry. This thinking was the basis for the revolutionary political strategy advocated by Mariátegui and his followers (see Romero and Levano, 1969; Delgado, 1982).

While 'democratic centralism' may appear to be a contradiction in terms, when democracy is defined from a class perspective one can begin to see how it may work in practice. Democracy is not simply the *process* of decision making and electoral politics. It is also the substantive question of which social groups and classes benefit from democracy – that is, democracy for whom? For *campesinos*, democratic centralism means liberation from an oppressive system of landownership supported by central authorities, and the establishment of a centralized system representing their interests and the interests of the oppressed classes in cities. This necessarily implies the political unity of workers and peasants. Democratic centralism implies a truly national state capable of bridging the gap between city and countryside because neither working class nor peasantry have a fundamental material interest (through the ownership of private property) in oppressing their ally. This does not mean that all contradictions between the two will disappear, or that problems between central authorities and rural areas, especially in underdeveloped countries, will end. It does mean, however, that the relation between urban and rural areas is no longer governed by an oppressive class system. In this sense the new centralism can be truly democratic.

Central planning may provide the conditions for decentralization, but does not necessarily lead in that direction. Primitive socialist accumulation can in fact produce greater centralization (as in the USSR in the 1930s); the central plan may increase inequalities, sometimes purposefully; and when planning follows an abstract ideal of equality without taking into account the laws of economic and urban development it can end up in disaster. Centralization that reinforces central elites and dilutes democratic participation may exacerbate inequalities. The examples below of China and Kampuchea seek to illustrate these points; the example of Cuba shows how democratic centralism can help eliminate inequalities.

Since central governments, and most decentralization schemes, have fallen far short of their goals, it is not difficult to understand why many turn from central solutions to local initiative as a remedy. It is logical to conclude that, lacking a redistributive central power, the only way inequalities can be rectified is through concerted and direct action at the local level. One alternative is to strengthen elective local government; another is to reinforce spontaneous self-help initiatives.

The colonial administrations in Africa, Asia and Latin America tended to be highly centralized. The colonial powers often destroyed popular, democratic and communal structures so as to enhance the role of their appointed regimes. After independence, many countries adopted the old colonial structures of local government. There is not always a lot of choice in the matter. Less developed countries have a small national surplus to distribute in the form of local services, so there is often no pressing need for local government, or a way to finance it. At the same time, a strong central government is often a necessary first step in a nation's attempt to control its own resources.

A decentralization strategy that rests mainly on promoting local government misplaces priorities, and often produces the opposite of democratic participation. The United States Agency for International Development is fond of projecting the U.S. experience of local federated government as a model for developing countries. They finance programs to enhance the revenue-generating capacity of local governments and train local administrators. Noting the lack of resources in developing nations, they treat local governments as if they were profit-making enterprises whose success must be measured by revenues and fiscal rationale. But what interests do these local governments serve? The interests of local elites and property owners, or the interests of workers and peasants? Do the local governments rationalize a central bureaucracy and overcentralized power structure, or do they truly decentralize power? Does strengthening local government mean cutting back national support for local programs as it does in Ronald Reagan's 'new federalism'?

If local governments are based on structurally unequal development, they only reproduce that pattern. Because their powers are geographically limited, they are unable to affect the overall structure of inequality even if they can more equally distribute the surplus available to them within their own boundaries. More often than not, the strategy of strengthening local governments is a camouflage for weakening them.

Self-help

Another favorite strategy promoted by the United States aid establishment is local self-help. The basic strategy is to finance local groups and organizations which arise spontaneously to deal with the miserable conditions in the metropolis, particularly in squatter settlements, and rural areas (see Turner and Fichter, 1972). The World Bank's Sites and Services program (World Bank, 1974) arose from the self-help strategy.

Support for self-help purportedly encourages the decentralization of power and resources. In reality, however, it rationalizes the process of unequal development. It can obscure both the inability and unwillingness of central governments to more equitably distribute national resources. It can obscure a mode of economic development that depends on maintaining a minimal urban infrastructure to maximize profits to capital.

Self-help housing programs promoted by the international aid establishment usually involve some form of credit and 'cost recovery.' This usually means that the housing consumers must have stable incomes in order to qualify for credit, which excludes people in the lowest income categories. It means housing consumers must pay back the banks and international donors for material assistance. This indebtedness in the end winds up promoting private property and a real estate market where only a weak one existed previously. Most housing in squatter settlements has little or no exchange value on the market. Mortgage credit, bank involvement and a modern infrastructure add up to a private real estate market able to house the upper strata but unable to solve the housing needs of the impoverished many.

This does not mean that self-help is not a positive element in development and should not be encouraged. Self-help actually describes the way most metropolises in developing nations have been built – spontaneously and without government assistance (see Hardoy, 1982). Sometimes it can be very organized and sophisticated. For example, in Chile under the Unidad Popular government, and in Mexico since the 1970s, local associations for improving the urban environment played key roles nationally as well as locally, and have been successful in pressuring for government financing. However, these groups rely on self-help out of necessity, not because they believe it is a preferred national strategy. Indeed, a constant among the strongest and most influential self-help groups is the demand for a stronger government role and a more equitable economic and political system.

Planned equalization

Decentralization within the context of a planned economy may involve growth poles, capital city relocation, self-help and local government, as in more market-oriented economies. However, developing socialist nations have tended to justify decentralization as part of a comprehensive national economic strategy that focuses mainly on eliminating dependency and structural inequalities between urban and rural areas.

In general, planned economies have shown more success at limiting the rate of growth of large metropolitan areas and promoting the development of small and medium-sized settlements. Forbes and Thrift (1987: 6) state that a slower rate of urban growth and reduction of the rate of primacy tend to characterize developing socialist countries. The authors compare urban growth in twenty-one socialist developing countries, and show how Cuba and Vietnam have had the slowest rates of growth and Tanzania, Mozambique and Libya have the highest. These variations are partially explained by the differences in the pre-socialist levels of urbanization (Cuba was highly urbanized at the time of the 1959 revolution, and Tanzania, Mozambique and Libya were mostly rural).

The question of whether these countries have altered the legacy of unequal development is a more difficult one; urban primacy is not necessarily an indicator of underdevelopment, nor its alleviation an indicator of development. In general, any evaluation of central planning must be qualified because socialism in developing countries is still, historically speaking, quite a recent phenomenon. Furthermore, while the approach of building a nation without private property in the main means of production is widely accepted in developing nations, it is practiced in so many different ways that generalizations are suspect. In former colonial nations, the state is often the only institution capable of owning and operating property abandoned by the colonial rulers, but state ownership by itself does not necessarily fulfill the objectives of socialism. Some nations whose governments consider themselves socialist have mixtures of private, state, collective and communal property. It is therefore not surprising that the results are usually mixed.

In the following, we will look at two of the more dramatic examples, and attempt to draw some tentative conclusions from them.

China's communes

The most important experiment with national settlement planning in former colonial nations was the commune system in China. It was based on Mao Zedong's revolutionary strategy, in which the countryside would surround and conquer the cities, and his vision of socialism, which was basically agrarian and anti-urban. Mao suspected urban elites of being 'capitalist-roaders' who were to be exiled to hard labor in the countryside. He projected the rural communes as self-sufficient production and distribution entities. Uneven development was considered entirely avoidable. Large cities and metropolitan areas were to be contained and migration limited. Economic development in rural areas would be on a small scale and prevent urban development. Mao spoke of 'the five smalls': fertilizers, iron and steel, hydroelectric power, agricultural machinery and repair, and cement. With this limited technology, the countryside would become industrialized without promoting urbanization (see Gurley, 1976).

Mao's theories were additionally embellished by the French Marxist Charles Bettelheim. Bettelheim considered the process of 'disurbanization' during the Cultural Revolution to be part of a more general struggle for direct control by the Chinese working class over the planning process. He saw this as a way of revolutionizing the relations of production and thwarting the reproduction of the old capitalist relations (Bettelheim, 1974; 1975; 1976; 1978).

Mao's communes were eliminated in the 1980s, after the end of the 'Great Proletarian Cultural Revolution' and Mao's death. Today, the commune experiment is considered by Chinese communists to have been a Utopian exercise based on egalitarian and not socialist principles. Recent policies have begun to reverse the previous ones (Gibelli, 1987; Kwok, 1987).

Thirty years after the 1949 revolution, about 84 per cent of China's population is still non-metropolitan. Though the gap between the standard of living in urban and rural areas has narrowed, it is still substantial enough to spur migration to the metropolis. The growth rate of China's cities has remained above 3 per cent, about twice the overall population growth rate. Wu (1987) shows how investment levels in cities have exceeded those in communes, despite the stated policy of favoring the rural communes.

Mao and Bettelheim reflected a common anti-urban bias and suspicion of the metropolis. They did not accept that development is always uneven, and uneven development does not necessarily reinforce structural inequalities. In fact, the extreme decentralization of the commune system ultimately reinforced the differences between city and countryside – because there are objective limits to a society's ability to provide a full range of goods and services to small communities. The communes simply could not equal the standard of living in the metropolis no matter how developed they became.

There are universal laws of economic development that function under both capitalism and socialism: the law of scale economies, the law of supply and demand, and the law of value. The law of value, which governs all economic laws in capitalism, operates partially under systems based on social ownership. The problem for planners is to acknowledge these laws so that they can be harnessed and controlled. Equality is not the central goal of socialism, which is a transitional and far from an ideal system; rather, the main goal is the appropriation of the social surplus by the workers and peasants, and distribution according to work. Egalitarian and Utopian schemes such as the communes hinder the process of economic development and therefore limit the ability to gradually redistribute resources and to prevent inequalities from becoming structural. Contrary to Mao, developing the productive forces *is* key, especially in the dependent nations, but it must not be forgotten that the relations of production – including relations between workers and managers, and democratic participation – are essential elements of the productive forces.

The most dramatic attempt to implement the Maoist anti-urban strategy was in Kampuchea under the Khmer Rouge (see Caldwell, 1977). Under the four-year reign of terror of Pol Pot, all cities were evacuated by force, and millions were sent to labor in the fields. The capital city of Phnom Penh went from two and a half million to 50,000 people. All signs of urban culture were considered counter-revolutionary according to the messianic vision of primitive communism engendered by the Khmer Rouge. Intellectuals, cultural workers and religious people were persecuted. Everything foreign was seen as undermining the Khmer Rouge vision of agrarian nationalism. The anti-urban policy was part of a plan to destroy everything associated with modern civilization, a plan that had genocidal consequences. As many as three million people, almost a fourth of the country's population, died at the hands of the Khmer Rouge.

Zimbabwe and Mozambique attempted milder versions of the anti-urban spatial strategy, and also failed. Vietnam had partial success in reversing the concentration of population in Ho Chi Minh City after liberation, but in the 1980s accepted the metropolis as an historical fact to be managed and developed.

Cuba

At the time of the victory of the Cuban revolution in 1959, Cuba was one of the most urbanized nations in Latin America, itself the most urbanized continent in the Third World. Over 20 per cent of the population lived in the capital, La Habana, a primate city that dwarfed the provincial capitals in size, economic importance and standard of living.

Within the first two years of the Cuban revolution, the most open signs of intra-urban inequalities were done away with. Shantytown dwellers were given new housing. The first urban reform laws stopped displacement by effectively halting evictions and speculation in urban land.

In over thirty years, Cuba has carried out an urban–rural equalization process through its centrally planned system. Priority was given to new industrial activities in Cienfuegos, Holguín, Santiago and other areas outside the capital. Health and educational facilities were dispersed to towns and villages, so that today Cuba ranks among the most advanced countries in the world in terms of life expectancy and proportion of doctors and technicians in the population. Cuba also has one of the lowest illiteracy rates in the world. As a result of the dramatic improvement of conditions in the countryside, migration to La Habana declined (see Susman, 1987).

Cuba is the only country in Latin America in which the primate city has stopped growing relative to the lesser cities. In fact, in the last thirty years, growth of the provincial capitals has outpaced growth of La Habana. In many ways, the level of services and quality of life in the capital have declined due to the priority given to rural areas and provincial capitals over the last thirty years. There are now serious problems in housing and urban services in La Habana that require new policy initiatives. These problems were probably unavoidable: correcting inequalities invariably means that areas once favored will be less favored *relative to the others*.

As noted previously, differences in settlement size, by themselves, tell little about economic and political equality. Still, enormous imbalances in city size tend to magnify uneven development because smaller settlements will never be large enough to enjoy a full array of social and cultural services, especially in less developed countries like Cuba. La Habana is one of Latin America's leading centers for film-making, publishing and art; despite a national policy of diffusing cultural opportunities to all rural areas through television, radio, the 'Cultural Houses' and financing of local talent, only La Habana affords the opportunity for the direct interaction among artists and intellectuals that makes such a rich cultural community possible. This is perhaps logical and necessary at this stage of Cuba's development. But it provides a challenge for the next stage of development, and speaks to the operation of certain laws of metropolitan development independently of central planning.

Although Cuba's centrally planned decentralization was responsible for progress in overcoming inequalities, spontaneous local initiative, including self-help, also played a role, and will probably play an

Figure 3.3 A Cuban microbrigade builds housing as part of a national strategy combining planning and self-help.

increasing role in the future. Since 1970, most new construction of housing and local community facilities was done by the microbrigade system, based on volunteer labor organized and backed by local enterprises and local and national government (Mathey, 1988). In addition, there is a substantial amount of self-built housing where state and microbrigade construction is not able to meet housing needs. Since inequalities in health and education are no longer substantial, access to housing and urban services is now one of the major indicators of urban–rural, and intra-urban, inequalities.

The questions facing Cuban planners in the country's next stages of development are increasingly complex. Up to this point, Cuba's planners have denied that internal markets can play any meaningful role in socialist economic development and the elimination of inequalities. This runs contrary to the direction taken by the republics of the former USSR and East European countries. It remains to be seen whether Cuba is able to make another leap in development without consumer markets. It appears, instead, that the objective laws of socialist development include internal markets, as they do reliance on central planning, self-help, local initiative and democratic participation. Cuba has made substantial progress in overcoming the legacy of the dependent metropolis. However, future progress in promoting equalized development will depend on a continuing ability to centrally plan not according to abstract ideas about equality but objective laws of development.

Since the collapse of the Soviet Union and decline of trade with Cuba by the former Soviet republics and East European nations, a number of new challenges face Cuban planners. It appears that Cuban planners in the past underestimated the significance of their dependency on favorable trade relations with the USSR and Eastern Europe, and the fragility of sugar-for-oil barters. Now they must face the alternative of trade on the international capitalist market, which means not only exporting commodities but also exporting surplus value to be converted into capital, and importing goods at higher prices. Since 1990, Cuban planners have identified three strategic industries for future development: sugar by-products, tourism (through joint ventures) and biotechnology. They now face the structural problem of all small Third World countries: without economic association with larger countries, they cannot have the productive base or market to produce a wide range of goods to meet the growing needs of the population (see Nkrumah, 1965).

Austerity measures caused by the decline in oil deliveries to Cuba from the former USSR have made life more difficult in both urban and rural areas. Perhaps the only silver lining on the cloud, from an urban planning perspective, is that auto use is no longer a viable transportation option, and bicycle manufacture and use in the cities has multiplied dramatically.

Conclusion: future prospects

The examples of China and Cuba illustrate the importance of national planning as a basis for equalized metropolitan development, and the failure of anti-urban approaches. The most successful urban planning strategies are based on successful national economic development strategies. The most successful national development strategies revolve around development of the human factor, not material aid or GNP, private property or profit. Development of the human factor means, concretely, concentrating resources on improvement of the daily living conditions of the population.

The prevailing trend in metropolitan development around the world is toward the continuing unplanned growth of the dependent metropolis, spurred by the expansion of transnational capital throughout Africa, Asia and Latin America. The future appears to promise more metropolitan areas and an aggravation of the inequalities between the metropolis and the rural hinterland, between the metropolis and cities, and within the metropolis.

Operating against this trend is another one toward more balanced development, in which metropolitan growth is limited and structural inequalities gradually diminish. The emergence of a more equal, decentralized urban structure requires, however, the adoption of equitable national economic and social planning. This is the route to truly 'sustainable development.'

It remains to be seen which of these two trends will predominate in the years ahead.

4

THE SOVIET METROPOLIS

> Property is theft.
>
> Pierre Joseph Proudhon

> Nothing that is new is perfect.
>
> Cicero

It is certainly a risky venture to attempt any comprehensive analysis of urbanization and planning in socialist countries at a time when the entire theory and practice of socialism is undergoing a revolution. Since the period of *perestroika* and *glasnost* in the Soviet Union in the mid-1980s, political revolutions have swept away the governments of Eastern Europe, and the USSR has dissolved. The ideological foundation of Soviet socialism, Marxism–Leninism, has all but vanished from the European scene, even if it continues to hold sway in many parts of the Third World, most notably in China, which has one-fourth of the world population. It is not clear what form socialism will take in the years to come; some doubt it will survive at all and others have already pronounced it dead. Yet so far social forms of ownership continue to dominate the main means of production in Eastern Europe and the republics of the former Soviet Union. Exultations by cold warriors in the West about the demise of socialism are premature, though it is certainly clear that socialism as it has existed since 1917 is already a thing of the past.

Developing socialist countries fare somewhat differently. They are in serious economic difficulties, and can no longer rely on the largesse of the Soviet Union. Capitalism appears to have gained new momentum around the world, particularly the welfare state European and Japanese versions, which are rapidly expanding in areas once considered off-limits to them. Yet for the majority of the developing world, including socialist and mixed economies, capitalism still offers very little of an alternative. Capital continues to invest mostly in the highly developed countries, and many Third World nations that have been in capitalism's orbit for decades have yet to experience the 'trickle-down' miracle.

The birth and death of modes of production take place over the course of entire epochs, in fits and starts, and not by single apocalyptic political acts. Ultimately, it is the history of the centuries following the revolutions of 1848, and not the years following 1917, that will provide the definitive answers to the questions about the transition to socialism.

No matter what the ultimate historical summation of them may be, the regimes of 'real socialism' that were born with the Bolshevik revolution produced a qualitatively distinct metropolitan form. In this chapter, we will describe and analyze the experiences of the Soviet Union, East European and other socialist countries in the realm of urbanization and planning through the 1980s, and try to show how these experiences have differed markedly from those of Western Europe and North America. We have discussed the experiences of developing socialist countries in the previous chapter. These countries have been influenced to some extent by the Soviet model, but many, especially China, have rejected that model. Their planning approaches have been driven by the search for alternatives to the legacy of colonial dependency – which was not the case in the Soviet Union and Eastern Europe.

In many ways, 'real socialism' was a pioneering project. It was innovative and also put into practice some of the most advanced ideas on city planning from the West. This much is a matter of historical record, and the record is important for understanding current changes and future prospects. This experience can and should inform tomorrow's planning, in socialist, mixed or capitalist economies. Without fully understanding this history, the current trends and problems of planning throughout the world cannot be adequately understood.

The Soviet experience was the first and has been the most influential among countries that have sought to build socialism. In this chapter we focus on the Soviet Union because the experience, planning approach and image of the Soviet metropolis have served as something of a model – both positive and negative – for socialist development around the world.

The demise of the USSR in 1991 provides a fitting opportunity for a broad historical analysis of the Soviet experience free from Cold

War prejudices. There will be a tendency in the West to use the opportunity to demonstrate the inferiority of socialism and futility of alternatives to capitalism. A more balanced approach will seek to identify both strengths and weaknesses, accomplishments and failures, in the Soviet experience.

Perestroika unleashed intense criticism of past theory and practice in economic and city planning. The Chernobyl disaster evoked widespread skepticism of technological determinism, and awakened consciousness of the environmental dangers overlooked by planners. Since the breakup of the USSR, local and republic governments have claimed new powers and called into question the old system of planning.

In city planning, there is now an emphasis on rehabilitation of housing over new construction and a greater emphasis on providing sorely needed community services. Private and cooperative enterprises are playing greater roles in urban design, development and management. There is a consensus that the old approach of mechanically reproducing neighborhood districts and housing units failed to take into account the quality of urban life. While urban pollution is a new concern, there are also new demands for individual consumption goods that could unleash a contradictory trend toward even greater environmental contamination.

In this chapter, we will first give an overview of urbanization in the Soviet Union and Eastern Europe, and the urban structure it produced. Then we will review the Soviet approach to city planning and comment on the problems and opportunities of the revolution unleashed by *perestroika.*

THE SOVIET METROPOLIS: AN OVERVIEW

Metropolitan development in socialist countries exhibits many of the same characteristics of metropolitan development in capitalist countries, an indication of the universality of the metropolitan form of development. However, there are also major differences between the capitalist and socialist metropolis which can be attributed to the mode of production, regime of accumulation, and the particular stage of economic development of the individual nations making up what was known as the socialist bloc of nations. Many urban problems of the capitalist countries have been largely absent from the socialist countries. But a whole new array of urban problems arose under socialism. Planning in the socialist countries has addressed these problems unevenly, and there are both successes and failures to point to.

Previous study by North Americans of urban planning in the Soviet Union and Eastern Europe has been scarce and often marked by ideological prejudice. In a widely cited article by Jack C. Fisher in the *Journal of the American Institute of Planners* (1962), the common prejudices about Soviet city planning were baldly expressed. These were criticized in Fisher, Pioro and Savic (1965).

Robert Osborne (1966) uncovered the fallacies common among Western observers. And Hans Blumenfeld (1978) critically analyzed the misconceptions of Larry Sawers (1977) about Soviet urbanization. Sawers tried to compare the Soviet Union and China and show that the Soviet Union was a more centralized country with an urban structure similar to that of the capitalist countries. He attributed the disparity to the ideological centralism of the Soviet planners and the decentralization policies of the Chinese planners. Blumenfeld argued that the facts show otherwise. China since 1949 had a higher rate of urban growth, and the percentage of the population living in urban areas was not substantially higher in the Soviet Union. This corresponds with the figures in Table 1.5, which show 16 per cent of China's population in metropolitan areas as compared to 19 per cent in the former USSR. Blumenfeld also reiterated his view that urbanization is linked more to the level of development of the productive forces than to the mode of production, and that central planning can influence aspects of the urbanization process.

Apart from these debates, a number of comprehensive studies available in English can help in developing a basic understanding of planning in socialist countries (Parker *et al.*, 1952; French and Hamilton, 1979; Bater, 1980; Musil, 1980).

In analyzing the Soviet metropolis, the main methodological problem for planners both inside and outside the socialist bloc has been physical determinism, a form of idealism. Socialist planners in particular have tended to equate physical form with political ideology. Western observers unconsciously mimic this approach by trying to read socialist urban forms as directly related to political theory or social structure. Both Western analysts and socialist planners tend to ignore the qualitative distinctions between city and metropolis that have characterized twentieth-century urban development on a world scale, including capitalist, socialist and mixed economies. In many discussions there is also an underlying Utopian conception of socialism and the socialist metropolis that is essentially anti-urban.

The Soviet metropolis had all of the salient characteristics of the modern metropolis as defined by Hans Blumenfeld: a relatively large territory and population, a complex division of functions, with central districts and suburbs, a central leadership function, and a high level of mobility (Blumenfeld, 1971). The differences are twofold: the relations among settlements in the national urban structure as a whole were distinct, based on a more evenly dispersed centrally planned allocation of resources; and the functional divisions and inequalities within the metropolis took on different forms and were not as sharp.

National urban structures

The urban structures of the former Soviet Union and Eastern Europe tend to be somewhat less dominated by large metropolitan areas than those in North America and the European capitalist countries. This is mostly due to objective factors: the particular historic and geographical conditions of Europe, the absence of a reserve army of labor, and a low rate of natural increase of the population in metropolitan areas. Political factors include national welfare policies, executed through central economic plans, that guaranteed basic services in rural areas, and national policies that promoted development of small and medium-sized cities.

Centrally-planned economies tend to have smaller metropolitan areas and a smaller proportion of the national population living in them. The average size of the metropolis in the former Soviet Union is about 1.8 million, about two-thirds the size of the average North American metropolis. Less than 20 per cent of the population is in metropolitan areas over one million. The proportion of population in medium-sized cities, over 100,000 and under one million population, is substantial, about 29 per cent (see Table 1.1).

The proportion of the USSR's total population in metropolitan areas did not change substantially since the 1920s. However, the number of metropolises increased from two in 1926 to ten in 1970 to thirty in 1990. Since the 1920s, the largest increase in urban population was in cities between 500,000 and one million population (see Blumenfeld, 1979: 185; Medvedkov, 1990).

The highest *levels* of urbanization in the USSR were in the European republics – Russia, Belorussia and the Ukraine. The highest *rates* of urbanization were in Kazakhstan and Central Asia (Kozhurin and Pogodin, 1981). Since the end of World War II, there was also a high rate of urban growth in southern parts of the Soviet Union due to the rate of natural increase and new industry; the south also started from a low base of urban population.

The average size of the East European metropolis is 1.5 million, compared to 2.7 million in Western Europe. In Eastern Europe, only 10 per cent of the population lives in metropolitan areas, compared to 29 per cent in Western Europe (although the proportion for Western Europe is probably underestimated).

The most dramatic contrast in size and urban structure is between the former Soviet Union and North America (see Chapter 1). The differences between Eastern and Western Europe are not as significant. There are several reasons for the greater similarity between both parts of Europe. First of all, the East European countries adopted centrally planned economies and eliminated the reserve army of labor only some forty years ago, thirty years after the Soviet Union. At the time they began implementing central planning, some of them were already industrialized nations with industrial cities, shaped by a common European history. Another factor that has brought East and West together is the growth of the welfare state in Western Europe after the war. The regime of capitalist accumulation wedded to a welfare state and historically powerful institutions of civil society tends to minimize the most extreme poverty of the reserve army of labor, and thus excess metropolitan growth.

The former GDR gives us an example of a national settlement structure based on limited urbanization through central planning. The GDR had only one metropolis (Berlin), which was really the lesser part of the greater Berlin metropolis. East Berlin was about one-third the size it was before World War II, and one-third the size of West Berlin. The rest of the GDR's settlement structure included two medium-sized cities – Dresden and Leipzig – and many small settlements dispersed throughout the countryside. Only about 14 per cent of the GDR's national population was in cities over 100,000 population. As Blumenfeld has shown, the fastest growing settlements in the post-war period were in the 20,000–50,000 population category, and not the larger cities (Blumenfeld, 1979: 182–3).

This settlement pattern was the result of the GDR's guaranteed employment policy and national planning that dispersed industries throughout the country. War reconstruction and subsequent development gave priority to towns and medium-sized cities. After World War II, many of the largest settlements in the country had been virtually destroyed. About a third of all housing in Berlin and half of all housing in the GDR was lost. Dresden lost 85 per cent of its buildings. Many towns and cities were destroyed, including Halberstadt, Nordhausen, Frankfurt/Oder, Prenzlau, Pasewalk, Anklam, Neubrandenburg, Potsdam, Plauen and Karl-Marx-Stadt (Chemnitz) (Bach, 1974: 450).

With the reunification of Germany, the urban structure in eastern Germany is already undergoing a rapid change. There is sizeable unemployment as the economy undergoes restructuring, and it is not clear how much of this will make up part of a more permanent reserve army of labor. There is dramatic migration to the West, but there is also substantial migration from other parts of Europe to eastern Germany. In any case, there is an unmistakable trend towards greater concentration of capital and labor in the major urban centers, and this is bound to alter the trend that evolved in the GDR over the last forty years.

Czechoslovakia before 1989 also had a relatively decentralized urban structure. From the beginning of its industrialization before the war, Czechoslovakia's industry was dispersed among many medium-sized settlements. In the 1970s, the average distance between settlements was only twenty-six kilometers (about fifteen miles) (Venys and Kohout, 1975). Today, only 19 per cent of the national population is in cities larger than 100,000 population and 11 per cent in the single metropolis, Prague. Prague's growth rate is not substantially greater than the average for the country as a whole.

Some previously rural East European countries urbanized rapidly after the war in conjunction with the industrialization of their economies. Poland went from 27 per cent to 60 per cent urban and Bulgaria from 25 per cent to 64 per cent (World Marxist Review Working Group, 1985). These figures indicate a high rate of urban growth, but they rely on broad definitions of 'urban,' because Poland and Bulgaria today have only 18.9 per cent and 17.3 per cent of their respective populations in metropolitan areas. Through central economic planning, welfare levels in small towns and cities were kept close to levels in metropolitan areas, and investment was dispersed in the less developed regions. In Poland, 85 per cent of agricultural land remained in private hands, and the flourishing of private agriculture brought greater wealth to many rural areas. At the same time, large state investments and relatively high salaries in the Silesian coal region produced a dispersed urbanization pattern that affected a large part of southern Poland.

The most important element in limiting growth has not been the conscious decentralization policies of urban planners. Such efforts have rarely succeeded by themselves. Instead, limited metropolitan growth has been related to the basic economic logic of socialism in the form it took since 1917. Most important are the national guarantees of employment, health and educational services. Thus, a special issue of *Studies in Comparative Communism* (1979) showed how socialist governments tend to give a higher priority to social services and education than to infrastructure for production. Also, without the reserve army of labor as an economic necessity, there was no demand for a surplus population in metropolitan areas. Wages and prices were established on a national basis, and differences in the quality of life between the metropolises, cities and towns were not as wide as in the capitalist countries (though in some countries they were substantial, and in others conditions were sometimes better in rural areas).

In the Soviet Union the policy of building new cities of limited size did not significantly contribute to the pattern of limited metropolitan growth. The new cities accounted for a very small proportion of total urban growth. Since 1917, about a thousand new cities, all of them under a million population, were built either around small villages or from scratch. Some of the most important ones were in the eastern part of the Soviet Union, in Siberia and the Far East, where large sums from the national budget were invested to tap the natural resources of the region and settle isolated areas. However, most were in the European parts of the Soviet Union, which were already the most urbanized.

The new Soviet settlements were planned for populations ranging from 50,000–250,000, and rarely were any planned at the scale of the metropolis. The largest, an exception to the rule, is Togliattigrad, in Russia, originally planned for a maximum 500,000 population, but now expected to reach at least one and a half million by the end of the century. Togliattigrad, named after Palmiro Togliatti, leader of the Italian Communist Party, was planned around a giant auto factory, perhaps one of the largest in the world, which was built under an agreement with the Fiat corporation.

New city building in the USSR was part of a national policy to disperse industrial development. The objectives were to promote development in areas with unexploited natural resources, particularly in the Far East, to correct the historical inequalities among the fifteen republics and 100 nationalities making up the Soviet Union, and to make the country less vulnerable militarily. During World War II, the Soviets picked up practically their entire industrial infrastructure and moved it east of the Urals to protect it from the invading Nazi troops.

After the war, central planners tried to divert most new industrial investment to towns and medium-sized cities. According to Morris Zeitlin, 'Most of the 3,200 new industrial plants built in the last decade . . . are in small and middle cities of urban agglomerations' (Zeitlin, 1985).

One of the enduring debates in the Soviet Union was over optimal city size. Despite the inability of planners to limit the size of Moscow and other metropolitan areas, the concept of optimal city size maintained a certain degree of credibility. For a long time it was official policy to limit the growth of new industry in urban areas over 500,000 population (see Stepanov, 1974: 1154). In practice, however, industrial enterprises and ministries proved to be more powerful than local and national-level planners, and new industrial firms managed to spring up in the metropolitan areas, where they could take advantage of linkages with other industries and the developed infrastructure. The growing service sector, whose natural habitat is the metropolis, followed suit.

These experiences show how the law of the metropolis functions within centrally planned economies as well as capitalist economies. The laws of scale economies, supply and demand, and locational value also operate, and policies that do not take them into account are destined to failure.

Contrary to the theory of optimal city size, the question of settlement size alone is not particularly significant in the quest for well-planned settlements. The most important question for planners is: are size differences indicative of structurally unequal economic and social development among settlements? More thorough analysis of the Soviet case, without ideological prejudice, is needed. Despite some glaring deficiencies, many towns and cities in the Soviet Union benefited from new housing and services. Many entirely new cities were relatively well equipped with basic health and educational facilities. There were always jobs in these settlements, because employment was constitutionally guaranteed to all regardless of place of residence.

However, the reality is that while social welfare policies tended to mitigate inequalities, the amount and quality of services available in larger cities and metropolitan areas was generally superior. The differences between the industrially developed European republics and the southern and central Asian republics are often substantial. It remains to be seen whether the breakup of the Soviet Union will reinforce or reverse this pattern.

The internal structure of the metropolis

Even tourists could spot some of the superficial differences between the internal structure of the U.S. and Soviet metropolis. In the Soviet metropolis, they found no Central Business District oriented around commercial offices and retailing. While they could see a seedy side in every urban area, they could not find an entire Times Square district with flashing neon lights, open drug dealing or homeless people. They could see that low-cost mass transit predominates over the private automobile (although traffic congestion has begun to plague metropolitan areas). They found more of the older historic districts preserved instead of redeveloped for modern office buildings.

Casual visitors would also lament the drabness of unpainted exteriors in many older areas, and the monotony of high-rise apartment blocks in newer areas. They would cringe at the sheer size and monumentality of suburban development. In the winter, they gasped at the shroud of coal dust that hovered over many cities in Eastern Europe, product of the crude fuel used for domestic heating.

These images tell us some important things about the Soviet metropolis, but they are only superficial and obscure many other important elements. For example, exterior drabness is not a very good indicator of overall housing quality, because it may conceal interior richness. It could indicate anything from a paint shortage to structural disrepair. And what may be aesthetically unpleasing to the tourist or visiting architect could work well as a community; what may seem revolting on the surface could be useful and functional.

These images obscure fundamental differences between the U.S. and Soviet metropolis. Segregation and inequalities among residential districts are much more pronounced in the U.S. metropolis; the Central Business District is clearly the main focus of urban life. The Soviet metropolis, on the other hand, has relatively uniform residential districts and a more varied land use in central districts.

The monotonous uniformity of residential districts in the Soviet metropolis, so often criticized by Westerners, was partially due to the mechanical industrialization of city building, but was also related to social homogeneity. The extremes of Park Avenue condominiums and Harlem tenements simply did not exist. If they did they would have been obvious even to the casual observer, and could not be masked even by dishonest social scientists. This is not to say that poverty and social differences were not serious problems in Soviet metropolises, but to assert that they did not generally take the form of spatially segregated districts. Social differences were rooted in wage differentials and the privileges of bureaucracy rather than ownership of property. While social privilege may take the form of more residential space, this space was generally scattered across the exurban countryside in the form of *dachas* and not concentrated in silk-stocking urban neighborhoods.

There was no private real estate market in the Soviet metropolis. Because there was no economic competition for land through a real estate market, land values in central areas did not increase, and these areas were not Manhattanized. Consequently, displacement was not a significant phenomenon. When households were displaced by urban renewal they were usually moved to new housing of at least comparable size and quality.

The relative uniformity of residential densities was one indication of the absence of a real estate market and dramatic differences in housing quality. In some metropolitan areas, such as Moscow, for example, the central density increased minimally over the years. The overall pattern of relative homogeneity of Moscow districts did not change significantly since the 1930s.

In the Soviet metropolis, almost all housing was socially owned – by the state, state-owned enterprises or cooperatives. In socialist countries where housing is privately owned, as in Cuba, the land remains under state ownership and the accumulation of real estate holdings for speculative purposes is not permitted. Rents tend to be uniformly low. In the Soviet Union, the average household spent 4–5 per cent of their income on rent (Marquit, 1983: 122). In most socialist countries, the proportion does not exceed 10 per cent. All of this is now changing, as local governments seek to raise housing payments to generate revenues for maintenance and rehabilitation, or to privatize housing and land.

Where there is no private real estate market, land does not have an exchange value. However, it would be a mistake to say that land has no value. It has a social use value. This value varies from one district to another, and one parcel to another. The use value of the land in the center of the metropolis is often higher than other parts of the region because of its locational advantage. This is a function of geography and not market economics, even though markets incorporate geography. Thus, while there was no Central Business District in the Soviet metropolis, there were centralized functions, which included retailing and government offices. The central areas were still magnets for cultural life. Many well-preserved historic centers (distinct from commercialized tourist districts like New Orleans' Vieux Carré, for example) continue to play unique roles in the metropolis.

Both visitors and residents would agree that the Soviet metropolis was no paradise. It had many of the same problems as the free-market capitalist metropolis, and a few others as well. There may have been no Times Square, but poverty, drugs, prostitution and crime were not insignificant phenomena. Mass transit may have been plentiful and cheap (a ride on the Moscow subway in the 1980s cost five kopeks, about a nickel, the same as in 1935), but as car ownership began to soar traffic congestion and air pollution continued to grow. The production system was wracked with inefficiency and waste, and industry was responsible for many environmental disasters. Social and retail services may have been planned along with every new residential neighborhood, but there was often a chronic lag in services due to bureaucratic mismanagement. Indeed, as the metropolis turned to services over industry in the post-industrial era, the old centrally planned system proved incapable of delivering adequate services to meet rising demands. The old central planning apparatus could not cope with the modern metropolis.

PRINCIPLES AND PRACTICE OF SOVIET CITY PLANNING

In the decade after the Bolshevik revolution of 1917, a widespread cultural revolution ignited many new theories of architecture and city planning. These theories remain to this day among the most influential, and debated, in the world. The Bolshevik revolution held for many intellectuals the potential of creating a new world based on the cooperation of labor instead of class exploitation, a world in which the latest technological advances would be used to improve the conditions of the working class. Freedom from private property promised to unleash freedom for the human race to shape its environment in a rational, scientific way, and to foster social welfare instead of private profit.

The basic theoretical foundation for city planning was considered to be Marx's classical critique of the division between city and countryside, which he attributed to the process of capitalist accumulation. In *The Communist Manifesto*, Marx and Engels asserted:

> The bourgeoisie has subjected the country to the rule of the towns. . . . The bourgeoisie keeps more and more doing away

> with the scattered state of the population, of the means of production, and of property. It has agglomerated population, centralized means of production, and has concentrated property in a few hands.
>
> (Marx and Engels, 1968: 39)

Among the things the proletariat would undertake on seizing power would be the:

> gradual abolition of the distinction between town and country, by a more equable distribution of the population over the country.
>
> (Marx and Engels, 1968: 53)

One of the earliest works of Friedrich Engels was a detailed analysis of the living conditions of the English working class in cities (Engels, 1973). In Volume I of *Capital*, Marx established a link between the accumulation of capital, the accumulation of labor and population, and the accumulation of misery (Marx, 1967). The socialist revolution would remove capital's pull of population to large cities, eliminate surplus labor and do away with the miserable living conditions of the working class. In *The Development of Capitalism in Russia*, Lenin had also documented the rapid growth of large cities in the latter part of the nineteenth century.

After the victory of the Bolshevik revolution, the possibility of eliminating the differences between city and countryside and ending urban misery appeared to be very real ones. A group of architects and planners who came to be known as 'disurbanists' proposed the planned dispersal of population outside the large cities. They proposed the establishment of many small planned towns across the countryside, somewhat along the lines of the British Garden City movement. For example, Moisei Ginzburg and Mikhail Barshch proposed a plan for a 'Green City' around Moscow in 1930. Ginzburg was one of the founders, with Alexander and Leonid Vesnin, of the Union of Contemporary Architects (OSA) in 1925, an association that advanced many of the principles that in the West became associated with Le Corbusier and the modern school of architecture.

One of the most important variants of the deurbanization model was by Nikolai A. Miliutin, who advanced a proposal for linear cities built along transportation lines, a proposal somewhat similar to the *Ciudad Lineal* of Spain's Antonio Soria y Mata and the *Cité Industrielle* of France's Tony Garnier. Miliutin and his associates proposed 'institutions for collective feeding of the population, collective education of children, as well as mechanized laundries – these are the first necessary elements of collective life that must be provided for in new construction' (Miliutin, 1974: 76). Miliutin adopted some of the ideas of the more extreme disurbanists and, along with Ginzburg, was one of the leading proponents of the rationalist trend in architecture known as constructivism.

Important new concepts were generated in the studies and debates of the 1920s. There was the concept of 'social containers' – urban places where a new and diverse social interaction was supposed to develop. These containers were to be cultural and recreational centers distinct from the alienating social environment of capitalism. There was also the concept of urban development based on a minimal, basic unit of communal life, first advanced by Leonard Sabsovich and Ivan Leonidov. This was further developed by S.G. Strumilin into the idea of a micro-region (*mikrorayon*), which has been the basic planning unit used by Soviet planners. A *mikrorayon* is about 5,000–15,000 population and a *rayon* is about 100,000.

The ferment in Moscow in the 1920s attracted many architects and planners from abroad. Leading architects from the West, including Le Corbusier and Frank Lloyd Wright, were attracted by the new ideas and the prospect of seeing entire communities built according to plan, a rare occurrence in the West. They absorbed some of the revolutionary ferment, engaged in exchanges and debates, and participated in competitions. In Moscow, there was a natural affinity with modernism as it was developing in the German Bauhaus, although there was an explicit critique and rejection of the modernist projects based on commercial profit.

The Great Turn and the 1930s: planning and primitive accumulation

Roberto Segre (1988: 439) notes that 'The objective conditions of economic and social development in the USSR in those years were not yet mature to put into practice the proposals of the avant-garde architects.' The Soviet economy was recovering from the destruction wrought during the civil war (1918–21). The New Economic Policy (NEP) was more oriented toward reestablishing rural production than establishing an urban life. Rural electrification was more important than the urban infrastructure. As a result, very little of the new thinking that arose during the 1920s was implemented in that decade.

After the 'Great Turn' of 1929, which marked the definitive move away from the NEP, conditions changed rapidly. The Soviet Union embarked on an intensive industrialization program, collectivization of agriculture, and the institution of central economic planning. The new economic policies created the conditions in which the ideas of the 1920s would be put into practice.

In the 1930s, the population in cities went from a fifth to almost a third of the national population. Moscow doubled in size. This accumulation of population in the cities and nascent metropolitan areas coincided with the fastest rate of industrial growth of any nation in the world.

The Soviet experience of the 1930s demonstrated for the first time that, independently of the disurbanist ideas of Soviet planners, the process of metropolitan development would accompany socialist accumulation as it had accompanied capitalist accumulation. If under capitalism the metropolis grows in accordance with the accumulation of private capital, which in turn leads to the accumulation of labor, under socialism the metropolis grows in accordance with the accumulation of social capital, which also requires the accumulation of labor. Private capital's surplus from production includes a portion that is not returned to labor, and is reapportioned by capitalists so as to further reproduce their own capital. Social capital's surplus from production has generally been accumulated centrally and reapportioned according to a central plan, both to reproduce social capital and to meet social needs. One of the problems of capitalist accumulation is that priority goes to reproducing capital and not to meeting social needs. One of the problems with socialist accumulation is that planners' priorities for reproducing social capital are often twisted; they have often favored basic industry, poorly interpreted social needs, and evaded the basic socialist principle of 'to each according to their work.'

The primitive accumulation of the 1930s, whereby the surplus from a modernized and rationalized agricultural production was applied to investment in heavy industries, led to the aggregation of the proletariat in urban areas. The new industrial plant was mostly located in cities and metropolitan areas, some of which were newly planned settlements. This, plus famine and repression of the kulaks in the countryside, spurred migration to the urban areas. And this occurred despite administrative attempts to prevent migration and limit urban growth.

Migration to metropolitan areas was constrained by other actions taken by Soviet planners. New industry was established in many towns and small cities. National social welfare policy guaranteed jobs and benefits in these smaller settlements. There was also a rural proletariat formed on large collectivized farms, the *kolkhozy*.

Elaboration of the first five-year economic plan began in 1928. In subsequent years the one-year program and ten- to twenty-year perspective plans were included in the repertory of central planning instruments. The Western powers, in the throes of the Great Depression during the 1930s, borrowed elements of Soviet planning in their efforts to bail themselves out of their predicament. Since there was little new building in the West during the Depression, opportunities to try out comprehensive planning in urban development were minimal.

At the threshold of the dynamic building boom of the 1930s, the Soviet leadership rejected the anti-urban ideas that arose in the 1920s. At the Bolshevik Party Congress in June 1931, the decentralist proposals for development outside the large cities were rejected as Utopian. The decision was made to emphasize urban development as an essential component of the ambitious industrialization plan. At about this time, the experiments with communal living arrangements were meeting resistance and soon came to an end.

Many of the innovative Soviet ideas of the 1920s were concentrated in the 1935 Moscow Master Plan, a seminal work of urban planning and a revolutionary project that influenced all future development in socialist countries, and capitalist countries as well. The Moscow plan integrated many of the avant-garde ideas current among planners in both capitalist and socialist countries. The scale of the urban infrastructure was ambitious, and the level of investment required for its realization astounding. It was an example of a monumentality that was to be reproduced throughout subsequent decades in the spacious layouts of superblocks and the imposing bulk of the buildings.

Many of the principles of the Moscow General Plan have proven to be durable components of urban planning practice around the world. They include the principle of integrated development, standard services distributed equally throughout the metropolis, the preservation of green areas, preferential treatment for mass transit, separation of heavy industry from residential areas, and the promotion of social interaction through physical form. There were illusions about the social impact of the plan based on Utopian and physical determinist ideas. But it was a bold assertion of the attempt to shape a new

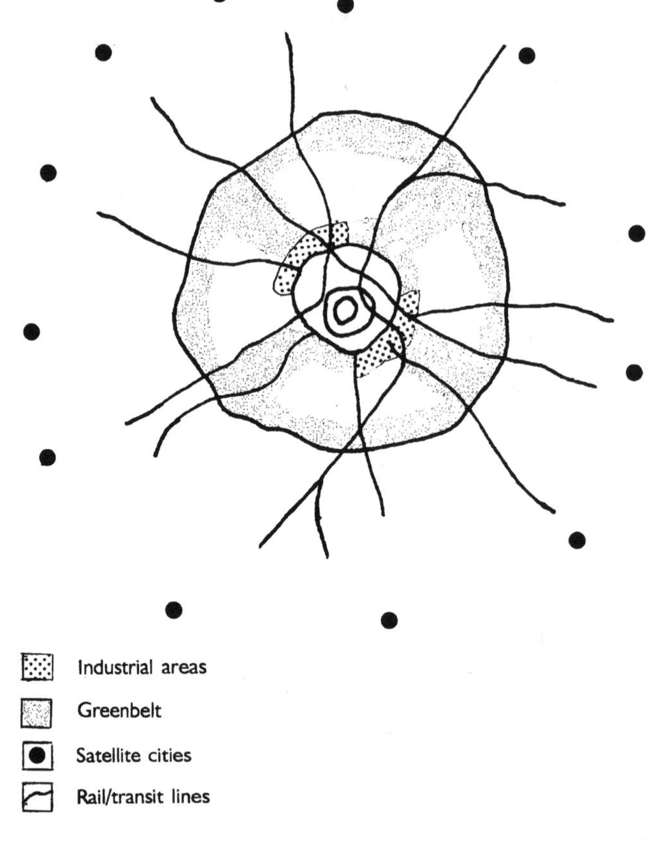

Figure 4.1 Schematic master plan for Moscow by S. Shestakov, developed in the 1920s. Shestakov's plan contains the elements that dominated subsequent Soviet plans: a greenbelt, industrial areas, satellite cities, and a rational transportation system made up of radials and concentric circles.

socialist society on the foundations of a modern industrial economy and central planning. Above all it was a truly comprehensive plan that linked and integrated sectoral plans.

The form of the future metropolis outlined in the 1935 plan followed the rationalist thinking prevalent in the 1920s. The districts within the metropolis were linked by a road and mass transit network in the form of concentric rings and radials, a geometry offering maximum mobility and efficiency of transportation. Along the radials a linear scheme was adopted similar to the one developed by Miliutin. The metropolitan area was to be encircled by a green belt that would mark the limits of urban development.

Most of the major projects incorporated in the plan were

implemented, many of them in an exceptionally short period of time. While the Stakhanovite movement promoting worker emulation accelerated industrial productivity, volunteer labor through the *subbotniks*[1] accelerated city building. The Moscow subway, for example, was built in a few short years, with substantial volunteer labor.

One aspect of the Moscow plan that did not prove valid was the limitation on the size of the metropolis. The plan projected a maximum five million population, based on the theory that there is an objective optimal size for urban settlements. This theory partially reacted to the capitalist experience of uncontrolled growth of industrial cities. The five million population level was exceeded after World War II, and the 1971 Moscow plan set eight million as a new ceiling. In the 1980s, Moscow passed the eight million mark, and the once inviolate green belt became the site of new urban development.

In the 1930s, the Soviet government for a short time attempted to limit the population of the largest cities and metropolitan areas by administratively restricting migration from rural areas. This policy did not work (it has never worked anywhere for very long) and was soon abandoned. The housing shortage in the cities and metropolises may have at times inhibited people from migrating, but the continued overcrowding in many Moscow apartments suggests that lack of housing does not serve as much of a deterrence to migration (this is equally as true in the capitalist world). Much more effective in slowing urban growth have been the policies of dispersed industrial investment and guaranteed employment.

The basic principles of Soviet urban planning were set out in the 1930s. According to Vladimir Semionov, they included the following ten principles:

1 Planned population limits are based on general demographic projections;
2 Communal services are planned within organically unified residential blocks, or 'superblocks';
3 The differences between city and countryside are to be eliminated by providing basic infrastructure to rural areas, not by creating new satellite cities;
4 Continuous regulation of individual projects is to insure conformance with urban plans;
5 The superblock is the basic unit of urban structure;
6 Cultural and political needs are included in the programs for community services;

7 The particular environmental, geographic, demographic and historical conditions are to be considered in establishing new urban areas;
8 National traditions should be basic components of architecture and planning;
9 The metropolis should be conceived as a unified whole and each action as part of a comprehensive approach;
10 In urban design, there should be consideration of maximum comfort for housing, development of the city center as the heart of the metropolis, the use of standards, the principles of socialist realism, and criticism and self-criticism.

(Adapted from Roberto Segre (1988: 465–6))

Soviet planning in the 1930s rejected the disurbanist approach, but Utopian ideas continued to play a role in Soviet planning. The theory of optimal city size, the layout of districts according to pre-cast models, and the reliance on ideological symbols were influenced by Utopian and idealist views. Although basically urban in orientation, Soviet planning theory failed to consider the metropolis a qualitatively distinct form of settlement. Therefore, planners underestimated the extent to which it would grow, and planned metropolitan areas the same way they planned small cities.

The more significant problem, that later came back to haunt Soviet planning, was a general tendency to impose ideological uniformity through political diktat in urban planning. In the early 1930s, Bolshevik leader Lazar Kaganovich led an attack on the Utopian ideas and experimentation of the 1920s that, while helping to get architects and technicians focused on the real, concrete problems of city building, managed to quash new ideas and experimentation. Kaganovich equated modernism in architecture with bourgeois culture, thereby rejecting one of the main theoretical foundations of all planning in the twentieth century and closing off Soviet theory and practice from the rest of the world. Debates in architecture and planning continued, and involved many of the same issues as the debates in the West (Osborne, 1966), but their scope was narrow – at least until *glasnost*.

Kaganovich reflected a simplistic and idealistic approach to culture which became all too common during the 1930s. This approach treated culture as a creation of political will rather than as a reflection of social reality. As a result, principles of architecture and city planning were viewed narrowly, as rational policies established by the political leadership; they were seen as a function of the political line of the Bolshevik Party and a budding proletarian culture.

At the time of the Bolshevik revolution, 80 per cent of the population were peasants tied to primitive agricultural production. In the 1930s, the proletariat became a majority, and its rise brought forth attempts to forge a new revolutionary, proletarian culture. This could not be done, however, with only a literacy campaign and slogans. The new culture would have to evolve over a long period of time as the young Soviet working class matured, and in the meantime it would have to be based on elements from national culture, including non-proletarian culture. Even then, the new culture, as with all culture, could not be simplistically defined only, or even mainly, in class terms, but would necessarily reflect all aspects of historical and social reality. For this reason, Soviet architecture and planning turned toward national cultures in the 1930s.

But there were also attempts to fashion a proletarian culture through central diktat and banal symbolism. There was the view that architecture and planning should always express in visual and aesthetic terms the values of socialism. This produced an emphasis on symbolism and monumentality that in practice could conflict with the socialist objective of producing a liveable environment for the new working class.

The practical tasks of city building in the 1930s, and later post-war reconstruction, undoubtedly overwhelmed Soviet architects. With the industrialization of the building industry, emphasis was placed on the wholesale application of already existing ideas and plans, on quantity over quality, so that the process of critical evaluation and dialectical renewal was thwarted. New ideas were not generated in the increasingly anti-intellectual atmosphere fostered by Joseph Stalin and the Party leadership. Criticism of Utopian ideas was carried out in a way that discouraged new thinking not directly oriented to production and immediate results. Soviet education focused on basic instruction and elimination of illiteracy, at which it was stunningly successful, but neglected the new frontiers of thought in many specialized areas. This was the beginning of the stagnation that endured into the 1980s.

Post-war development and stagnation

In the next stage of Soviet development, post-war reconstruction, the problems that first emerged in the 1930s multiplied. During the war, the Soviet Union lost over 1,700 cities and towns, and more than

70,000 villages were totally or almost totally destroyed. Twenty-five million people were homeless (World Marxist Review Study Group, 1985: 121). The job of urban reconstruction was an enormous one in a country badly fatigued from years of war.

Reconstruction required and encouraged a quantitative approach to urban development. The greatest priority was to shelter the largest number of people as quickly as possible. The goal of housing all homeless people was reached within a matter of a few years. But the intellectual environment in the country further degenerated and innovation was discouraged. This situation began to change after the death of Stalin in 1953, but the break with a mostly quantitative approach to city building did not seriously take hold until decades later.

As the Soviet economy developed, the economic structure of the metropolis began to change. In 1940, 44 per cent of Moscow's employment was in the industrial sector and by 1978 it was only 27 per cent (French, 1984: 375). However, the metropolis was still thought of as an industrial city. The goal in new cities was to replicate industrial concentrations; thus, a Soviet planner wrote in 1974, that about 20 per cent of the labor force in new towns was in the service sector, and in twenty-five to thirty years the proportion was expected to increase to only 40 per cent (Stepanov, 1974: 1147–8).

By the 1950s, the Soviet economy was among the most developed in the world. However, from the Great Turn in 1929 until *perestroika* in the 1980s, the nation's basic economic development policy and regime of accumulation did not change to meet the new challenges. Central planning degenerated into overcentralization and bureaucracy. The continued emphasis on heavy industry blocked the development of consumer goods. When attempts were made to mass produce consumer goods, like housing, the methods of production and distribution used in heavy industry were mechanically adapted without taking into consideration the variety of consumer needs and social practices. At the theoretical level, this reflected a denial of an independent social role for the individual, household, community and nationality.

In sum, the Soviet system was what has come to be known as an administrative/command system that ran on inertia instead of planning, and could not adequately respond to either popular needs or political leadership. Overcentralization in city planning discouraged the development of local initiatives and solutions adapted to local geographical and historical conditions. New construction was favored over rehabilitation because it was administratively easier. Uniformity in design was considered a more efficient option by bureaucrats. Bureaucracy thwarted the development of services in communities, especially new communities, producing the chronic problem of service lag. Local Soviets were hostage to the decisions of national ministries and a centralized Communist Party, and often unable to confront particular local problems where they were best suited to do so. Neighborhood and civic institutions that could have enriched the planning process did not exist or were paralyzed.

The post-war years of stagnation were not devoid of new ideas and experiments. There was significant experimentation with different building types, scales and decorative styles. In planning, new ideas on the creation of larger, modern urban systems emerged (see Gutnov *et al.*, 1968). The concept of 'urban group systems,' developed in the 1970s, explored ways to link and group together agglomerations and cities in a new settlement structure (Zeitlin, 1985; Medvedkov, 1990). Though the new theories did not explicitly acknowledge the qualitative distinction between metropolis and industrial city, this was a clear underlying premise. It was now more widely recognized that expansion of the large cities and existing settlements was inevitable and desirable.

In Moscow, planners evolved a view of the metropolis as a polycentric region with a qualitatively distinct metropolitan character. The decision to expand the satellite centers in the Moscow region to over a million population each as part of a more complex urban system was made in the 1960s. The 1971 Moscow Master Plan codified this approach. It started from the assumption that Moscow is a complex and diverse metropolitan region. Greater emphasis was placed on providing a variety of services within a variety of urban districts. Priority was given to improving regional transportation systems and protecting the urban environment.

The housing question

Perhaps the most serious urban problem in the Soviet Union and Eastern Europe over recent decades was housing. The main problem was a shortage of housing, but the shortage was very different from the shortage in capitalist countries. It was not a matter of high rents, displacement, evictions, tenant rights or speculation by landlords.

Housing was constitutionally guaranteed and provided at extremely low cost to all sectors of the population. The problem was basically a lack of housing units to take care of all households, overcrowding within existing units, and a mismatch between supply and demand. It was a matter of quantity, but was also intimately bound up with problems of quality.

On the demand side, the sources of the housing shortage were: vast war-time destruction of housing, the increased rate of divorce and growing proportion of single-person households, and the shortage inherited from the past. On the supply side they came down to the policy of favoring investments in industry over consumption, lack of units for households of varied sizes, and the quality of design and construction.

The housing shortage in the USSR was one of the most chronic and resistant to solution, and criticism of the housing shortage was on top of the list of criticisms that first emerged with *glasnost*. The problem was recognized by the Khruschev government in the 1950s as a serious one, and the goal was set of 'solving the housing problem' by providing a housing unit to every family by 1980. Investments in housing went up dramatically, and the Soviets built at the rate of almost two million units per year by the 1980s. In Moscow an average 100,000 units were built in the 1960s and 1970s; over half of all the apartments in Moscow today were built since 1956.

Between the mid-1950s and mid-1980s, 65 million dwelling units were built in the Soviet Union (Andrusz, 1987). Though the rate of new urban construction went up significantly in the 1970s and 1980s, the target date set by Khruschev for resolution of the housing crisis passed without his final goal having been met. Today there is an awareness of greater deficiencies in housing quality.

The housing shortage was most critical in metropolitan areas. Married couples often had to wait years for apartments, and single people even longer. Divorcees had nowhere to move. Three generations often shared a small living space. About 15 per cent of all units housed more than one family (World Marxist Review Study Group, 1985: 123–4). In Moscow there were an average 1.2 families per dwelling unit – a dramatic improvement from the 4.0 average of 1956 but the figure reflects the continuing lack of sufficient dwelling units (Kerblay, 1968: 52). The standard minimum living space in the Soviet Union was only 9 square meters per person in 1981, and the actual proportion in metropolitan areas was about 5 square meters per person. The standard was changed to 14.6 square meters per person in 1985, bringing it closer to the standard in the East European countries. This is still far below the standards in the developed capitalist countries, and about half the United States standard. The standards applied to new construction, which could only partially resolve the problem of overcrowding in existing units.

New construction tended to favor assembly line production of small one- and two-bedroom apartments in high rises, and the demand for both larger and smaller living spaces at a variety of building scales was usually not met. In Moscow, for example, about 56 per cent of new housing built in the 1960s was for three-person families, yet they accounted for only 27 per cent of the total (Kerblay, 1968: 75–6). This eventually changed, and there were renewed efforts to build more units for single-person households, including dormitory-style spaces with public eating places, and more larger units for large households.

The emphasis on new construction to meet the deficit in housing units further highlighted the problem of inadequate investment in rehabilitation and maintenance of existing housing. Despite increases in spending for housing rehabilitation during the 1980s, national budgets in socialist countries still favored new construction. While housing vacancy and abandonment were not major problems as in some capitalist countries, years of inadequate maintenance produced a substantial stock of housing in need of major repairs.

Maintenance is not a category amenable to quantitative reporting and results are not easily demonstrated to central planning bureaus and monitoring agencies, or to political leadership. It is much easier to claim as an accomplishment the construction of a thousand units of new housing than operation and maintenance of a thousand units. Also, maintenance is often left to centralized municipal agencies which have limited budgetary flexibility, and which tend to be removed from day-to-day management problems on individual blocks and in individual buildings.

Low rent payments are clearly an important social benefit. However, when there is substantial surplus money in circulation and little to buy with it, increasing tenant contributions to maintenance through local authorities may be one way of guaranteeing better services, and could be linked with increased tenant control. In the Soviet Union, rent covered only about one-third of maintenance costs. This amounted to an annual subsidy of over eight billion rubles. This did not include, of course, amortization of building costs.

Eastern Europe

Many of the accomplishments and problems of Soviet planning were reproduced in the East European socialist countries after 1948. However, over the last forty years there were also many independent developments in Eastern Europe.

In Poland, post-war reconstruction of Warsaw and other cities brought forth a new focus on the preservation of historic values. The reconstruction of Warsaw's medieval town square (*Glowne Rynek*) according to documents that had been preserved during the war was both a symbol of the nation's determination to reestablish its identity after the Nazi terror, and a unique experiment in city planning.

Poland's planners also experimented with regional planning by decentralizing planning authority to the voivodships (provinces). Perhaps the most successful experiment was the planning of an agglomeration of settlements including twenty-five new towns around the Silesian coal region in southern Poland. However, regional and local planning efforts were thwarted by the administrative/command system, and real power was never given to local bodies. Polish planners also experimented with mathematical models and optimization techniques to a greater extent than in the other socialist countries (see Fisher, 1966).

One need only look at Prague and other Czechoslovakian cities for examples, among the best, of preservation of historic central cities. Substantial public funds were spent on the preservation of housing as well as civic monuments. Strict regulations governing rehabilitation guaranteed a respect for historic typologies.

By contrast, most of the new development in Eastern Europe was more along the lines of Soviet rationalism and monumentality. Almost completely destroyed after the war, East Berlin was designed as a socialist model, with a giant tower that later looked down over *Die Mauer* (The Wall) to declare the superiority of socialism to a Western metropolis with slums and honky-tonk districts. Berlin's central square, Alexanderplatz, was surrounded by socially integrated worker housing in large modern apartment blocks. Public services and transportation were cheap and plentiful. As the wealthiest East European nation, the German Democratic Republic (GDR) had the highest standard of living, and Berlin was to show it off to the West.

However, Alexanderplatz also epitomizes the historic error of planning with symbols. It is a huge open area of concrete with only a small fountain and entrances to shops. Pedestrians are made to walk long distances, and to leave the superblock they must take the pedestrian tunnels (that really are anti-pedestrian) or cross the monumental avenues that ring the block. As car ownership grew, Alexanderplatz became increasingly engulfed in traffic and pollution. Since the annexation of East Berlin by the West, the monumental avenues now announce the dominance of a new auto-oriented metropolis. Unfortunately, the new Berlin may be unable to preserve the many progressive reforms of the West – reforms that could be attributed to its special status as an urban enclave. The new Berlin is already being flooded with autos, speculative capital, homeless people and a new reserve army of labor from the east.

Despite the problems symbolized in Alexanderplatz, the GDR was one of the prime examples of a balanced urban hierarchy made possible by central economic planning (as noted earlier in this chapter). Substantial investment went to small cities and towns, especially for industrial development in the southern part of the country. This contributed to an urban pattern in which metropolitan growth was limited, even more so than the relatively decentralized urban network in West Germany.

Romania is an example in the extreme of the problems of the administrative/command system in city planning. Since the 1970s, the Ceauşescu dictatorship undertook to redevelop the capital, Bucharest, once known as the Paris of the East, as a symbol of the new society. Its policy of 'systematization' entailed the destruction of old central neighborhoods. Bucharest is the only East European country where major sections of the historic central city were torn down, not as part of war reconstruction but as part of a vast urban renewal scheme. In their place, large modern buildings followed. The mammoth International Hotel, wide boulevards, and large apartment blocks contrast with the low-rise traditional buildings that once characterized the heart of the metropolis.

In the last year of the government of Nicolae Ceauşescu, which was overthrown in 1989, a proposal was launched to demolish 8,000 small villages on the theory that they were not economically viable. The residents, which might have included 50,000 ethnic Hungarians, would have been resettled by the government. While the displacement that occurred under Ceauşescu, and would have occurred had the Ceauşescu regime continued in power, did not leave people homeless, it is hard to imagine how it could not have had a disruptive effect on people's lives. Ceauşescu's administrative–bureaucratic approach did not consider such hardships to be an obstacle to urban development.

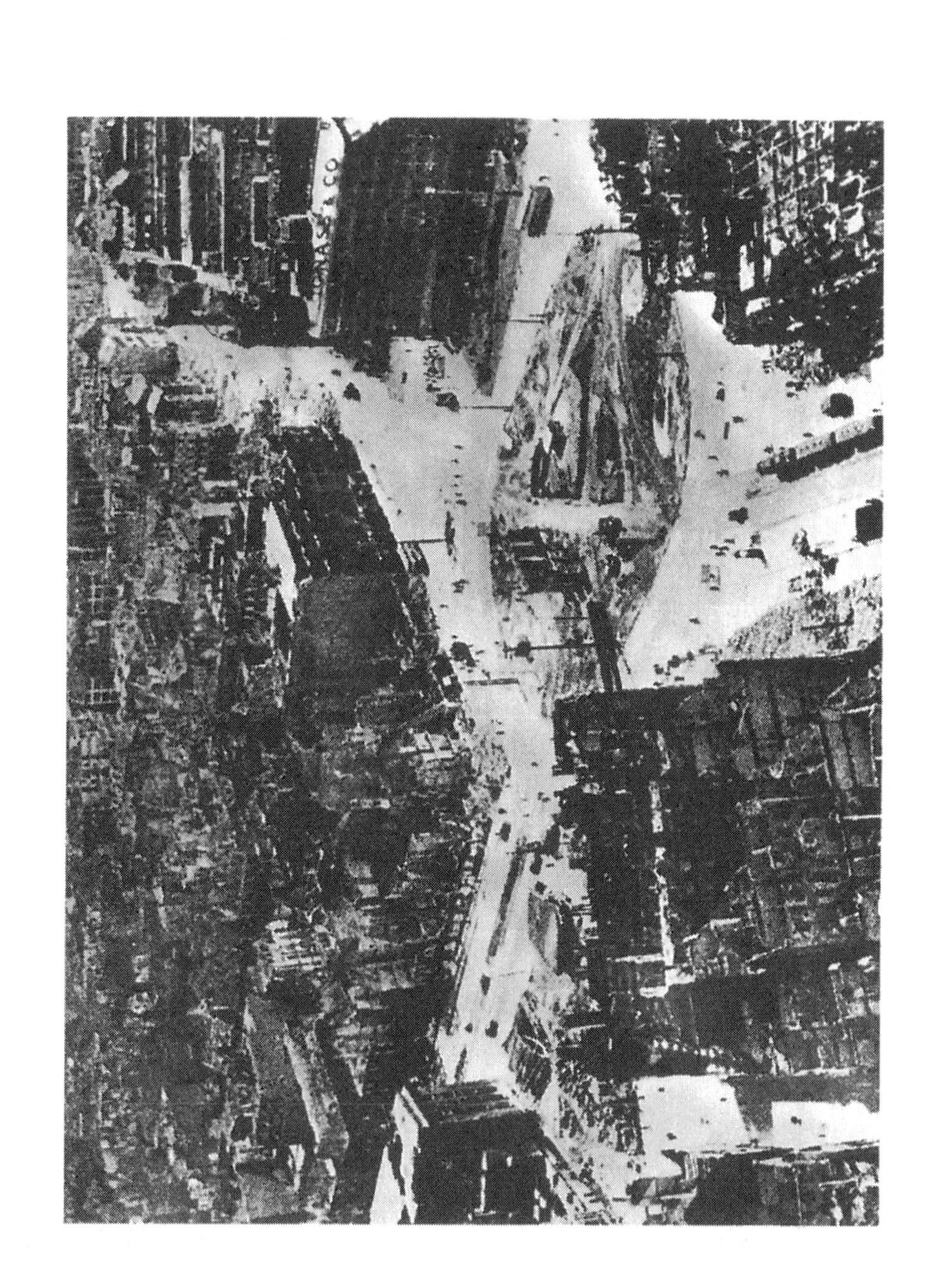

Figure 4.2 Alexanderplatz, East Berlin: (a) at the end of World War II; (b) after reconstruction.

Another of Ceauşescu's accomplishments was a grinding national austerity program which allowed Romania to pay back its foreign debt in record time.

After the Soviet Union

At this writing, *perestroika*, the Soviet Union and Soviet socialism are history, and it is not clear what will follow in the newly independent republics. The debates in city planning triggered by *perestroika* are still alive and have not been definitively resolved in new policies and strategies. The 'new way of thinking' called up some of the old ideas going back to the 1920s, and now the debates have intensified. It would be mere speculation at this point to predict where the new revolutionary ferment will lead, but at this point we can identify some of the main questions that have come into focus.

In the surge of criticism emerging with *glasnost*, there was renewed concern and debate about the housing shortage, the lack of rehabilitation and maintenance, monotony of scale and design, and the service lag. There were new proposals for alternative forms of city building and housing tenure, including decentralist schemes for dispersing the metropolitan population.

The revolutionary process unleashed contradictory trends. On the one hand, there is a new headlong thrust toward making more consumer goods of higher quality available to all. On the other hand, there is a rising concern over pollution caused by inefficient means of production and consumption. Yet if the experience of the West teaches us anything it is that a system that produces a variety of high quality consumer goods with relative efficiency produces more and not less waste and environmental damage. In the GDR before unification, there was one auto for every two persons. With unification, emission controls are bound to improve, but the total number of autos is likely to double, resulting in little if any improvement in air quality. With deindustrialization and the application of Western pollution control standards, industrial pollution may decline, but the growth in consumption may compensate the gains.

Thus the trend toward universal car ownership, as a means of improving individual consumption and evening out inequalities, contradicts the commitment to improve air quality in urban areas. It could well lead the former socialist bloc to replicate the auto-centered sprawled metropolitan areas that prevail in North America.

Within Soviet metropolises, there was in the past a naive and sometimes cavalier dismissal of problems related to the automobile. With renewed interest in the environment, planners can now take another look at the long-range environmental impact of the road designs that have dominated the urban landscape. The monumental avenues in urban areas, often with 200-foot roadways, are far wider than necessary for mass transit. The standard for Soviet planners was to use at least 25 per cent of urban land for streets and roads, a proportion comparable to that in the sprawled metropolitan areas of the United States. With proliferation of the private automobile, the wide avenues and streets of the metropolitan centers are readily becoming magnets for traffic.

There is now much discussion about the introduction of markets to distribute public goods, and strategies for removing some price controls. There are also many proposals for privatization of some or all public services. These include new forms of social ownership of housing other than state and enterprise ownership, such as cooperatives (in the USSR 80 per cent of housing was state-owned), and new forms of private ownership. While it is not yet clear to what extent the former Soviet nations will develop mixed, capitalist or socialist economies, and how they will alter the dominant regimes of accumulation, at the very least elements of markets are likely to be introduced. Prices will be increasingly used to regulate supply and demand of many public services and collective goods. There will probably be regulated markets in housing (but perhaps not real estate), food services and personal services. Some markets may be dominated by private enterprise, others by mixed public–private firms, and still others will be controlled by state monopolies.

A regulated market of limited-equity homes, cooperatives and condominia may well be able to respond more quickly to changing demand and offer greater mobility to the population, without the negative aspects of an unregulated private market. This could be achieved if land remains in social ownership.

The Soviet experience revealed how the problem of housing supply is bound up with problems of quality. The housing shortage was used as a convenient excuse for planning mistakes, monumentality, insensitive urban design and monotony. Especially in the years after war reconstruction, there was no reason other than stagnation in the planning field why the quantitative goals of housing production could not be met with a high quality product. Since the Bolshevik revolution, there was rarely a shortage of trained architects and planners, nor was there a real estate market to create a low-income housing shortage.

In the republics of the former USSR, there is now a new revolution in the theory and practice of urban planning comparable to the revolution of the 1920s. The center of gravity for the new revolution is located in cooperatives that have formed outside the realm of 'official architecture.' The cooperatives have contributed to the criticism of inertia, rigidity and extreme rationalism in city building. They have spearheaded a new thrust towards experimentation, creative freedom and pluralism in style (Chajcenko, 1989). Once again, planners are opening up to the international debates and dialogues in architecture and planning. The search for a new theoretical understanding of the laws of urbanization and planning is closely related to the striving of planners for a new relationship between theory and practice (Latoar, 1989).

The new revolution is coming to grips with the idealism and Utopian thinking that have plagued socialist planning from its inception. Idealist strains in the past have prevented planners from understanding that there are aspects of human interaction that at any given historical period cannot be fully understood or planned. The idea that everything can be governed by a central plan, through the administrative/command system or any other system, runs against the basic economic laws of socialism. As Engels once noted:

> The solution of the social problems, which as yet lay hidden in undeveloped economic conditions, the Utopians attempted to evolve out of the human brain. Society presented nothing but wrongs; to remove these was the task of reason. It was necessary, then, to discover a new and more perfect system of social order and to impose this upon society from without by propaganda, and, wherever it was possible, by the example of model experiments. These new social systems were foredoomed as Utopian; the more completely they were worked out in detail, the more they could not avoid drifting off into pure fantasies.
>
> (Marx and Engels, 1968: 403)

While the Soviets rejected classical Utopian planning, they engaged in the elaboration of highly detailed, rational, idealist plans that ultimately ran against the objectives of socialist development and objective economic laws. Consider the following:

> The reaction of the state power upon economic development

> can be of three kinds: it can run in the same direction, and then development is more rapid; it can oppose the line of development, in which case nowadays it will go to pieces in the long run in every great people; or it can prevent the economic development from proceeding along certain lines, and prescribe other lines. This case ultimately reduces itself to one of the two previous ones. But it is obvious that in cases two and three the political power can do great damage to the economic development and cause a great squandering of energy and material.
>
> (Marx and Engels, 1968: 696)

In sum, the idealism underlying the administrative/command approach runs counter to the dialectical and historical materialist method propounded by Marx:

> Men make their own history, but they do not make it just as they please; they do not make it under circumstances chosen by themselves, but under circumstances directly encountered, given and transmitted from the past.
>
> (Marx and Engels, 1968: 97)

INTRODUCTION TO CHAPTERS 5–7

In the last three chapters, we shall summarize the experiences of metropolitan governance and planning in the twentieth century, and suggest new approaches that might help avoid the pitfalls of the past and meet the challenges of the twenty-first century. At this point, it is important to note where our discussion is leading.

Contrary to free-market and anti-planning prejudices current in the United States and the current fascination with deregulation and privatization, the main problem has not been the presence or absence of planning but the quality of planning. Too often official planning is based on anti-urban prejudices which dictate that only a dose of the countryside can cure the ills of the city. Too often it is based on abstract Utopian models of the perfect city, and blind physical determinism. Too often it relies on statist notions that planning, especially good planning, can only be carried out through centralized government institutions.

Metropolitan-level planning must be the starting point for addressing the problems and opportunities of metropolitan areas. The problems of urban poverty, inequality and the environment require metro-wide solutions. An efficient physical and social infrastructure requires a strategic overview of metropolitan trends and development. Attempting to solve the problems of the metropolis without a metropolitan perspective is like trying to build a railway without surveying the land in between stations.

Perhaps the most serious problem with metropolitan planning practice today is the continued application of ideas and methods developed in the pre-metropolitan era of the industrial city. Metropolitan planning practice tends to be based on simplistic notions of master planning and master building that ignore the complex division of labor and functions characteristic of the modern metropolis. This has been the case in the former Soviet Union, Europe and throughout the developing nations of the Third World where formal metro-level planning tends to be most widespread.

A missing element in metropolitan planning and governance is *neighborhood-level empowerment*. Neighborhood planning is yet undeveloped in most metropolitan areas, and where it is most developed, in the U.S. metro, it is based on exclusion and segregation and not integration of functions and people at a regional level. Neighborhood planning is the key to integrating the household with the metropolis, residence and workplace. Women have played a pivotal role in pioneering neighborhood political action and community development, and it is no coincidence that the seminal works on neighborhood planning have been by women.

Metropolitan- and neighborhood-level planning need to be integrated through a new set of political and social institutions, a new synthesis of state and civil society geared toward a metropolis of *integrated diversity*. One problem is that the main forms of local governance around the world are leftovers from previous eras. National and municipal governments still wield most formal political power, yet the metropolitan elites and transnational corporations have gained significant economic powers. Metro-level governments, where they do exist, tend to be weak and controlled by national government, and neighborhood-level governance tends to be a mere appendage of municipal government.

A new progressive approach to planning is evolving. It is based on a vision of metropolitan development as a necessary accompaniment to social development. The metropolis should be seen as a public resource to be preserved for the use and benefit of all people, an inclusionary metro. It should be a place where people have maximum mobility without the threat of displacement, and where the greater social efficiency, equity and quality of life made possible in metropolitan areas may be maximized. Neighborhood-level political power and planning is central to progressive metro planning.

In the final chapter, we shall examine an array of strategies for neighborhood empowerment and planning. They include decentralized government, community economic development, and housing development and preservation. To be successful, these strategies require substantial support from higher levels of government and a restructuring of national priorities. Unless neighborhood strategies are tied to a more comprehensive process of metropolitan-wide planning, they can heighten inequalities between communities. Unless they reject the blind faith in growth and make community preservation a priority, local strategies will reproduce local problems.

There have been many attempts to integrate neighborhood and metropolitan planning by creating 'new towns in town' or 'cities within cities.' Some of these new communities have worked and some have not. Some promote exclusion and social isolation. Some are designed as separate enclaves with collections of private spaces. Some are cold monumental works of rationalism. However, there are many others that promote overall metropolitan integration, openess and public spaces. Many have become vibrant and urbane landscapes. These new communities have been built in market and centrally planned economies, and in developed and developing nations. Their success may be measured by the extent to which they adhere to the principle of *integrated diversity*.

5

THEORIES OF URBANIZATION AND PLANNING

> Anarchy, anarchy! Show me a greater evil!
> This is why cities tumble and the great houses rain down,
> This is what scatters armies!
>
> Sophocles, *Antigone*

Two prejudices consistently impede the evolution of a scientific approach to planning the metropolis: anti-urban bias and anti-planning bias. They are part of the Anglo-Saxon tradition of urban theory that underlies the growth and development of the U.S. metro, and influences metropolitan planning across the globe perhaps more than any other theoretical trend.

In the first chapter we discussed the anti-urban bias as applied to the analysis of world urbanization. In this chapter we will examine in greater depth the relationship between anti-urban bias and orthodox urban theory. We focus on the Chicago School of urban sociology as the fountain of orthodox theory. We will look at how the anti-urban bias intersects with Utopian theories.

We will also show how anti-planning bias is a pillar of contemporary urban theory in the United States, where rejection of the extreme rationalism common elsewhere in the world is often an excuse for the abdication of planning. It is part of the contradictory idealist philosophical framework which on the one hand explains away social problems as inevitable consequences of natural law that planners cannot change, and on the other sees social progress as a consequence of rational ideas instead of an objective development process. We seek a middle ground for planning practice that is based on reality and is able to change it.

Another fundamental bias, which affects North American planning in particular, is an anti-theoretical bias. True American pragmatism declares that theoretical discussions such as the one we are about to undertake can never be more than impotent exercises of ideas, or ideologies, and contribute little to planning because they do not offer direct and immediate solutions to urban problems. The pragmatist bias values only that which 'works' in the present. It goes along with a North American approach to economic and land-use planning that values project completion and short-run economic return over long-run utility and historic values. I will attempt to show how questions of theory are ultimately crucial in planning practice, contrary to the pragmatist bias.

Much of the focus here is on theoretical developments in the United States, which have widespread influence throughout a world in which the U.S. metropolis is held up as a model. The discussion leads to an appreciation of the efforts by a U.S. planner, Paul Davidoff, to fuse rationality, democratic participation and advocacy in the planning process. Davidoff's approach is 'proactive' and offers guidance for progressive planning in the complex modern metropolis.

ANTI-URBAN BIAS AND THEORY

As noted in Chapter 1, Lewis Mumford, a piercing and insightful critic of city planning, was nonetheless influenced by anti-urban prejudice. Mumford warned of the 'sprawling giantism' of the metropolis and counseled a return to settlements of more modest dimensions. Mumford more than many explicitly aimed his doomsday predictions at the metropolis. However, Mumford was expressing a more deeply rooted philosophical tendency within North American culture.

The fundamental assumption of the anti-urban bias is that urbanization causes urban problems. It is decreed that the metropolis breeds crime, violence and every form of social deviance. It is supposedly too 'urban' or too 'metropolitan' – too densely populated, too large, too diverse and too divorced from nature. Anti-urban bias has existed since the founding of the first cities. In the age of the metropolis, it has become an anti-metropolitan bias.

The anti-urban approach has led to efforts to bring the central city to the countryside – such as the Garden Cities and British New Towns – or to bring the countryside to the central city – such as the urban beautification movement in the United States (Howard: 1965; Osborn and Whittick, 1969; Scott, 1969; Odmann and Dahlberg, 1970). In both cases, the underlying assumption is that there is something drastically wrong with the metropolis that only a dose of the countryside can cure. Anti-urban planners also like to use medical metaphors that suggest that urban ills can be 'cured' by applying remedial measures (and even surgery, or excision of the 'cancer') (see Boyer, 1983: 9–32).

Anti-urban theories have little to show for all their efforts. Over ninety years after Ebenezer Howard launched the Garden Cities idea in England, the only concrete results are some chic suburban developments and token new communities. The urban reforms they have prompted tend to benefit the few. In the U.S., early experiments such as Sunnyside (Queens, New York), Radburn (New Jersey) and Greenbelt (Maryland), were never repeated. The government-sponsored new towns program folded in the 1970s after less than a decade, and the planned suburbs of Columbia (Maryland) and Reston (Virginia) are being surrounded by conventional suburban subdivisions. And all of Britain's New Towns house the equivalent of only 15 per cent of Greater London's population (see U.S. Department of Labor, 1971; Central Office of Information, 1972).

While in the United States the practical program of anti-urban theory was never implemented in any significant way, its basic assumptions have had a profound effect on the planning profession. The anti-urban bias appeals to the most conservative social impulses – a return to small-town living, narrow rural values and the presumed bliss of single-family homeownership. It idealizes the detached single-family home on a separate lot as the urban 'American dream.' It romanticizes what Karl Marx once called 'the idiocy of rural life' in a country that has little rural life left.

As a philosophy, anti-urbanism recalls many of the archaic ideas that ruled in pre-capitalist societies. As politics, it tends to buttress conservative policies ranging from fiscal cutbacks in social services to exclusionary zoning. Ruth Glass (1989) shows how these ideas promote gloomy pictures of public housing to justify its neglect, and proposals for authoritative social engineering. Especially as the overall needs of capital dictate declining expenditures on urban amenities, more and more planners are jumping on the 'less is better' and 'small is beautiful' bandwagon (see Schumacher, 1973; Meadows *et al.*, 1974; Sale, 1985).

Utopias

The ability to envision a new and better urban future is an important element in planning consciousness. Throughout history, Utopian dreams of better futures have contributed to developing this consciousness. But they have usually started from an anti-urban premise.

The term Utopia, which literally means 'no place,' was used to name the ideal construction of society put forth by Sir Thomas More in 1516. Other Utopias were Plato's *Republic*, Thomas Campanella's *City of the Sun* and Edward Bellamy's *Looking Backward*. The proposals of Count de Saint Simon, Charles Fourier and Robert Owen arose as alternatives to industrial capitalism in the nineteenth century and projected the possibility of a new society based on the cooperation of labor. Their ideas produced actual experiments in new communities – mostly rural – but none lasted for very long. They have been followed by a long array of intentional communities in the twentieth century based on many different religious, political and social philosophies. These communities have had brief life cycles, remained small, or have been swallowed up by the metropolis (Kanter, 1972). All of the ideas and practices they generated, however, are part of the modern critique of metropolitan society and planning.

Utopian ideas flourished in the early years of the Bolshevik revolution, when many of Russia's leading architects projected an urban future of dispersal across the countryside rather than concentration in metropolitan areas. Utopian socialism has also had its nostalgics, such as Mao Zedong, Pol Pot and Peru's Shining Path, who have attempted to go backwards to the ideal of primitive communism and consider everything urban to be potentially counter-revolutionary. There have also been right-wing nostalgics, like the fascist founders of Colonia Dignidad in Chile (which was disbanded after the return of democracy in 1990).

Utopian schemes have mostly looked backward toward the reestablishment of rural settlements rather than forward toward a metropolitan future. Fortunately, the return of society to bucolic austerity is precluded in practice by the objective laws of population and metropolitan growth that operate independently of the will of individuals.

Some Utopian visions look more toward the future. These include B.F. Skinner's *Walden II* and other small-scale experiments in social engineering. The more ambitious futurist experiments such as Paolo Soleri's Arcosante in Scottsdale, Arizona, and the recent Biosphere, may not be anti-urban in concept but their experimental environments typically end up in isolated ex-urban environments, accessible to only the few.

There is no place for Utopia to go in the metropolis. The modern metropolis is particularly unsuited to Utopian experiments. Its very size and complexity defy the imposition of simplistic models of social development and formalistic models of spatial development. Therefore, many modern-day Utopias tend to enter the realm of futurism, a clearly speculative one with little possibility or pretension of implementation in the present.

The Utopian impulse is present in contemporary idealized notions of the neighborhood and community. These ideas often challenge the integrating tendencies of the metropolis. In the United States, this is related to worship of the ethnic enclave, home rule and municipal authority, and opposition to regional and national formations. It usually translates into, or covers up, exclusionary and discriminatory practices. In Europe, it has recently taken the form of rising regionalism and nationalism, in opposition to immigrant populations living in metropolitan areas and to the process of European integration, which makes greater mobility within Europe possible (see Keating, 1988; Rudolph and Thompson, 1989).

One corollary of the anti-urban thesis is that existing metropolises are destined to die a natural death because they have outlived their usefulness. This thesis is indebted to *Social Darwinism* which, when applied to urban theory sees urban systems and sub-systems obeying natural laws similar to those that govern biological life cycles. In the United States, it is common to hear the view that there is a natural tendency toward urban decline and the only thing left is to let central cities die gracefully (U.S. Congress, 1977). This doomsday philosophy is almost always applied to the poor and working-class neighborhoods that are being forced out of existence, and *not* the Central Business District or new gentrified neighborhoods, which presumably are the fittest for survival.

The Social Darwinist philosophy gets translated into public policy in a number of ways. For example, former New York City Housing Administrator Roger Starr advocated a policy of 'planned shrinkage,' which would accelerate the decline of poor neighborhoods by cutting off their services. This is justified by the presumed inevitability of the decline (Starr, 1976). Another patron of this trend is Daniel Patrick Moynihan, a liberal Democrat with a reputation for intellectual enlightenment, now a senator from New York. Moynihan was President Richard Nixon's urban advisor when he promoted the policy of 'benign neglect' of the demands of the civil rights movement. This neglect, which has not been benign at all, had the effect of removing public support from African-American and minority communities. In practice, it left minority communities prey to the ravages of the 'natural' process of urban development, but did not fundamentally diminish the plentiful government support in the form of tax benefits and infrastructure subsidies to white communities.

Classical anti-urbanism: the Chicago School

The classical version of the anti-urban approach was formulated by the Chicago School of urban sociology, which originated at the University of Chicago in the early part of the century. This is probably the most influential body of urban theory in the United States. The Chicago School's approach is based on the notion that urban settlements are relatively independent social entities – independent of the laws of economic development – and have certain universal characteristics or functions. Urban problems are considered to be an inherent part of urbanization.

The Chicago School is significant because it was the first concerted intellectual attempt in the West to come to grips with the revolutionary transformations that occurred with the rise of the modern metropolis. The subject of the Chicago School was usually the metropolis, especially the North American metro (see D. Smith, 1988).

There have been several useful critiques of the Chicago School. These are summarized in Castells (1977), Gottdeiner (1985), Logan and Molotch (1987), and elsewhere. In the following, I will not try to duplicate previous critiques but review some of the key premises of the Chicago School as a means of uncovering the methodological problems. The Chicago School was extremely diverse and made up of many different voices, and some of its proponents were less anti-urban than others. We shall focus here on a common underlying thread and not the divergencies.

Louis Wirth set forth the Chicago School definition of a city in his classic 1938 article:

> A city may be defined as a relatively large, dense, and permanent settlement of socially heterogeneous individuals.
>
> (Wirth, 1964: 66)

Wirth did not really explain urban development in historical terms, or as a reflection of economic and technological development. Instead, urbanism was 'a way of life.' The urban way of life was juxtaposed with the rural way of life, and all societies were located on a continuum between these two ideal types:

> Thus, the larger, the more densely populated, and the more heterogeneous a community, the more accentuated the characteristics associated with urbanism will be.
>
> (Wirth, 1964: 68)

Wirth considered the 'urban way of life' the epitome of 'civilization,' and urban problems necessary evils. He and others in the Chicago School considered these problems to be inherent features of urbanism, ultimately related to the three salient characteristics of the urban settlement – size, density and heterogeneity. In other words, urban problems come about because settlements are big, dense and socially heterogeneous. The essential features of urban settlements are given as both the reason for their 'civilization' and the source of their problems. For the Chicago School, the ultimate determinant of urban problems is the metropolis itself, and not the socioeconomic system within which it exists.

Of these three characteristics, perhaps the only one that can stand up to an objective test is size, but it is also the one that tells us the least about the metropolis. As shown in Chapter 1, the qualitative increase in the size of the metropolis over the industrial city is one of the most readily identifiable aspects of urbanization in the twentieth century. Yet as shown in Chapters 2–4, the metropolis has taken on vastly different shape in different economic systems and regions of the world due to inequalities in the international division of labor.

Density differences between industrial city and metropolis are not very significant. There is evidence that densities in many modern metropolitan areas are no greater than densities in earlier settlements, even pre-industrial settlements. For example, one of the earliest settlements in history, the city of Ur, probably had a density of 6,300 people per square kilometer, about the same as the density of the Brussels metropolitan area today. Bogotá, Colombia, had a density of about 2,300 persons per square kilometer in 1670, about the same as Cologne today. Densities in pre-industrial cities ranged from 200 to 5,000 persons per square kilometer, about the same as the range of densities in the sprawled metropolitan areas of the United States today. (See Chapter 1 for a discussion of density and other examples.)

As for heterogeneity, there are numerous examples of metropolitan areas around the world that do not have the degree of social diversity the Chicago School found in the segregated metropolitan areas of

North America. One need only look at the relatively homogeneous metropolises of Sweden, Germany and China, for example.

It was no accident that the Chicago School reached a peak of popularity in the 1930s, when the metropolis was becoming a consolidated feature of capitalism in the United States, and massive worker upheavals were occurring in the nation's largest factories and urban areas. It was easy enough to attribute urban problems to the metropolis instead of linking them with the most severe crisis of capitalism heretofore known to the world, the Great Depression. If there were overcrowding, slums and homelessness, it was because urban settlements were too large and dense. If there was racism and discrimination against immigrants, it could be chalked up to the urban way of life and the American melting-pot. Such were the practical political implications of Wirth and the Chicago School.

The myth of the melting-pot in the United States is a favorite of popular sociology that mirrors the approach of orthodox functionalism in sociology. The melting-pot myth assumes that an equilibrium evolves among competing social groups through successive waves of integration. The fact that some social groups never melt, and most melt only partially, is either ignored or seen as part of another natural law which forever guarantees a permanent 'underclass' or culture of poverty (see Glazer and Moynihan, 1970; see Leacock, 1971, for critiques of the culture of poverty thesis).

The Chicago School was an integral part of the emerging tradition of functionalism in modern sociology. This trend, led by Talcott Parsons, quite consciously set out to carve up social reality into many isolated elements, each of them the subject of a separate social science discipline. The fragmentation of analysis yielded fragmentation of action (Mills, 1959). Parsons' social science menu was viewed as a comprehensive alternative to Marxism, which in the 1930s enjoyed considerable prestige in the working-class movements, as the Soviet Union's dramatic economic growth offered a sharp contrast to the crisis in the West.

The Chicago School introduced a form of what may be called urban determinism by making the metropolis itself the central, determining factor in urban development. The functionalists could take umbrage in the fact that Marxists were unable to fully explain the particularity of the metropolis, unravel all its internal laws of motion, and identify the extent to which it was a determining factor in social reality. Marxists often reduced urban problems to economic ones. Still, the functionalists got lost in their own narrow empirical observations of urban life. In the long run, their positivist approach has left out so much of reality that they have been unable to explain even their limited data.

The Chicago School's methodology was very often static. Society was not seen in motion, as a constantly changing process. In reality, social formations, and their settlement systems, are continuously transforming themselves, and these changes are both quantitative and qualitative in character. The static Chicago School approach liquidated the importance of abrupt, qualitative changes in urbanization and tended to freeze reality in a still photograph where only gradual, quantitative changes can occur. The Chicago School's ideal-type theories established *a priori* categories – i.e., ideal types of settlements – and then attempted to fit all urban reality into them, instead of starting with social reality and then setting up and modifying the categories in order to explain it.

Robert Redfield, one of the Chicago School's prodigies, was among the first twentieth-century social scientists to apply the ideal-type method to urbanization. He postulated two ideal types of society: the rural, traditional society and the urban, modern society. This rural–urban dichotomy was presumably to explain the differences between life in rural Mexico and metropolitan Mexico City, or between life in the towns of the Ozarks and Chicago. Every settlement could simply be measured on one continuous scale extending from the ideal rural type to the ideal urban type. Most supposedly lie somewhere between the two extremes (Redfield, 1947; Miner, 1952).

In static ideal-type theories, change is not entirely absent. However, it is depicted as the re-positioning of phenomena within static categories, without taking into account the transformation of these categories themselves. All reality becomes a scramble of isolated phenomena in a bipolar world.

A dialectical approach, on the other hand, seeks an organic, unified body of scientific theory in order to explain the world, and a comprehensive strategy to change it in an evolutionary and revolutionary way. It attempts to identify the actual motion of urban development in history, distinguishing quantitative and qualitative, and gradual and abrupt, changes. It attempts to trace the actual development and transformation of particular urban forms, of metropolitan areas and systems of metropolitan areas, and explain the laws that govern these changes.

Within the static framework of the Chicago School, a group of scholars posited a more dynamic process by examining urban settlements as part of evolving ecological systems (see Theodorson, 1961). Robert Park, Ernest Burgess and R.D. McKenzie (1925) stated the classical ecological position. Ernest Burgess (1961), attempted to show how urban settlements grow outward in concentric circles through a process of succession of land uses and social groups. Homer Hoyt proved this model too rigid, and showed how urban growth could follow a sectoral pattern radiating from the center; Chauncy Harris and Edward Ullman illustrated a pattern of multiple nuclei (see Johnson, 1972: 170–82). The urban ecologists, while searching for a more dynamic theory, have nonetheless wound up with models that assume natural urban processes, similar to biological systems, and rigid universal urban structures.

The Chicago School failed to adequately demonstrate the qualitative distinction between the industrial city and the metropolis. In their static world, the metropolis was simply the result of a long evolution of settlements toward an ideal. The fact that the ideal urban settlement was typically North American revealed the Chicago School's national and cultural bias.

Aside from being static, ideal-type theories are, of course, a form of idealism. The idealist method, which sees all social development as springing from the human will, looks at urbanization as a manifestation of abstract constructs rather than concrete economic activity. For example, Louis Wirth's theory of 'urbanization as a way of life' tries to explain urbanization as a complex of ideas, attitudes, habits and customs, rather than as part of the objective process of production and reproduction of the means of subsistence, the foundation of all human activity. Wirth sees only part of urban reality, the ideological and psycho-social aspects of urban life (and those only partially). (See Castells (1977) for a further critique; also Gans (1968) for an early analysis of 'suburbanism as a way of life,' which challenged Wirth's notion of a hegemonic urban way of life.)

Theories of urbanization that see the economic basis of society as simply one among several determining factors try to avoid the pitfalls of economic determinism, but in the process obscure the determining role of economics. As a result, subjective factors assume a disproportionate weight in these analytical models. Factors such as culture and ideology, key components of the 'urban way of life,' assume greater weight than the division of labor and the reproduction of capital and labor. Thus, the Chicago School and the general anti-urban bias that it reflects lead to an incomplete understanding of the metropolis.

Planning models that are based on these idealist and incomplete explanations of the metropolis, if implemented, produce planning failures. Utopian visions of society leave out the real economic laws that underly all social development, and therefore falter in practice. Ideal communities from Plato to B.F. Skinner have had much longer subjective lives than objective ones. But even where these ideals are not implemented in a comprehensive way, they tend to have a profound influence on more practical planning models, in which anti-urban and idealist approaches to the metropolis often play a significant role. Allen J. Scott (1980: 236ff.), for one, acknowledges these 'idealist–utopian origins of mainstream planning theory.'

ANTI-PLANNING BIAS AND THEORY

Some anti-urbanists who rail against economic determinism rely on the most blindly deterministic economic theory – that which attributes absolute determination in urban development to the interplay of market forces and denies any significant present or future role to conscious planning and intervention in the human environment. This leads them to conservatism in planning, an ideological hybrid combining the anti-urban bias with an anti-planning bias.

For example, Edward Banfield's (1970) famous theories of social austerity are based on the idea that urban poverty is a relatively minor social defect which the metropolis has inherited and cannot resolve. Banfield believes that there is no reason for government intervention in central cities because the urban marketplace will automatically correct the correctable problems, and the others, like poverty, will always be with us. The poor are poor because of their own inability to get out of poverty, because they are born into the culture of poverty. In his study of poverty in southern Italy (which might as well be Harlem), Banfield finds the culprit to be the 'amoral familism' of the Italian peasantry – a cultural weakness condemning them to perpetual poverty – while he conveniently overlooks such factors as the dominance of northern Italian capital, the North's drain on the South's resources (especially labor), and the powerful rule of the Christian Democratic Party, the Church and the Mafia. Banfield's 'benign neglect' blames the victims of poverty for poverty (Banfield, 1958).

This type of theory assumes the existence of absolute natural laws of two kinds: natural laws of social development and natural laws of political decisionmaking. All planners need to do, supposedly, is to tune in to these natural laws, and make sure that no one interferes with them. The order of the day is: let the economic marketplace have a free hand, unfettered by plans, and let the political marketplace go its way, unfettered by advocates of social change. It is hard to conceive of a more pessimistic approach to planning, and life.

Charles Lindblom's classical essay, 'The science of muddling through' (1959) put faith in the logic of the decisionmaking marketplace rather than the logic of democratic human consciousness. His pragmatic goal is short-range incremental changes rather than long-range planning. It leaves for planners the role of mediating among competing property interests, working out compromises to get the maximum benefit for all concerned with the minimum loss. This kind of cost–benefit compromising requires that planners simply follow the 'invisible hand' of the urban marketplace and not interfere with its 'self-regulating mechanisms.' These are not, strictly speaking, planners, but might better be considered facilitators, administrators, or just plain bureaucrats.

As a guide for the multiple institutions engaged in the planning process, both inside and outside official planning institutions, marketplace theories mandate conservatism and loyalty to the status quo. They start with the assumption that planners can or should be independent from political and institutional power, go on to show how they have never been able to achieve independence, and end by counseling relative passivity.

Advocates of theories that worship the spontaneity of the marketplace undercut attempts at planning, and then point to the failure of such efforts as proof of their theses. These theories immobilize any attempt to change the metropolis – either its social, economic and physical shape or the political forces that rule it.

In general, social theories that posit natural, immutable laws of development impute eternal features to social systems that are, in history, only short-lived and which, in the broad scope of human development, have not proven themselves eternal at all. Within this framework, socialist development or mixed systems cannot really be explained in any serious way. Instead, we have many superficial attempts to show how these 'natural laws' operate even in economies that have amputated the invisible hand at the wrist and dispensed with private property, the cornerstone of the capitalist marketplace. These attempts fail to develop a theory of the state, state legitimation or the legitimation crisis.

This is not to say that markets and private property do not, or should not, operate in centrally planned systems. Markets can function without private property in the means of production, to distribute consumer goods and allocate social resources. There is no reason why limited land markets cannot coexist with comprehensive economic and urban planning. (The 1989 revolutions in Eastern Europe and the demise of the USSR have not necessarily brought about a victory of the capitalist marketplace, either in the means of production or in real estate, but may produce a new mixture of public and private ownership, marketplace and planning.)

The dichotomy between planning and marketplace does not get at the heart of the problem of planning in the twentieth century. It is only a surface contradiction. The central problem has to do with the question of whose interests are served by planning and the market. *Who plans and who benefits?* Free-market theories obscure these questions, even in those cases where markets may serve the interest of the greatest number of people; they tend to obscure fundamental property relations and issues of social equality. (See Hernando de Soto (1989), who glorifies the unregulated informal economy in developing countries by considering it strategic to economic development.)

Despite a drastic turn in recent decades toward market theories as part of the privatization and fiscal austerity promoted by the Reagan/Thatcher revolution, the role of the state throughout the world has not diminished. In general, austerity has simply increased the rate of capital accumulation by cutting the proportion of the surplus returned to labor via the state, and increasing returns to capital. Thus, while the Reagan/Thatcher revolution has tried to bury the Keynesian revolution, it has merely buried it in the myths of early capitalism. Reagan's budgets were the highest ever, and state support for capital greater than ever before. Of course, much of the growth in state spending, and in government planning, was for military purposes (see Adams, 1984).

Physical determinism

Instead of diminishing the role of planning, physical determinism inflates planning beyond its real possibilities. Physical determinism is a unique philosophical problem. It is the main occupational hazard of the architect and urban planner. Since these professions are necessarily concerned with the concrete impact of physical form on social function, they tend to overemphasize this impact (see Gans, 1968: 25–33).

There is always a tendency in the design professions to ignore the social and economic context of planning and fetishize the formal aspects of design. The epitome of success is achievement of a unique design – in less noble professions known as a gimmick – to which unusual powers may be attributed. One may even fabricate a whole school and call it post-something-or-other based on minimal design innovation. Professional associations are especially concerned with promoting such notions, not only to nurture the reputations of a few master builders but to boost the status, and sometimes the business prospects, of the entire profession.

In his analysis of contemporary architectural theory, Tony Schuman considers two prevailing models: architecture as a 'lamp' that lights the way 'toward a nobler, more satisfying world through the force of the built and visual environment'; and architecture as a 'mirror' that 'implies a conscious design decision by an architect to use built form as a commentary on society, to reflect its strength or palliate its weakness.' Modernism is associated with the lamp and post-Modernism and deconstructivism with the mirror. Both project an heroic role for the profession and exaggerate the social effect of the built form, especially the individual building. Schuman proposes instead a more modest, socially responsible 'framework for existence' that deals with the everyday problems of people, and 'identifies the practices and rituals of daily life, mundane and extraordinary, as the subject of architecture' (Schuman, 1991).

An opposite tendency, to underestimate the significance of the physical environment in shaping human consciousness and social function, is manifest in a tendency toward economic determinism. It arises principally outside the design professions, and stems from a failure to grapple with concrete social relations as they interact with, shape, and are shaped by the physical environment, i.e. the spatial dimensions of social relations. It is a different form of abstract idealism that posits economic relations only in one-dimensional non-spatial terms.

Physical determinism not only ascribes wondrous virtues to planning, but also attributes every imaginable evil to poor planning, poor design or, more generally, a poor living environment. This type of physical determinism can even coincide with, or masquerade as, social consciousness. It meshes with the myth of the planning technocrat as problem solver for social ills.

The greater the possibilities for planning, the greater the inclination toward physical determinism. Planners in centrally planned economies have been particularly susceptible to this malady. The monumentality in much socialist city building is testimony to the determinist outlook.

We may trace the origins of modern city planning to the European reform movements of the nineteenth century, particularly in England and France. With the defeat of the 1848 revolutions, philanthropists sought to address urban squalor through model housing schemes, such as the worker houses built during London's Great Exhibition of 1851, and the 1867 Paris exhibition. These experiments were also accompanied by urban housing legislation. However, governments devoted their greatest resources to physically imposing large-scale public works:

> during the twenty years following the 1848 Revolution the first large-scale planning operations were carried out in the cities of Europe – the *grands travaux* of Haussmann in Paris (1853–69), and of Anspach in Brussels (1867–71), the building of the Ringstrasse in Vienna (1857 onwards), the development of Barcelona (1859 onwards) and Florence (1864–77), the re-designing and installation of a main drainage system in London where, between 1848 and 1865, Joseph Bazalgette completed the new system of main sewers along the Thames, the Victoria and Albert Embankment, and where the Underground was begun in 1863. These developments were the work of a new class of planners and civil servants, scientific, competent, and satisfied with their various departmental responsibilities.
>
> (Benevolo, 1967: 110)

The model for physical determinism in the United States was the Columbian Exposition at the 1893 Chicago World's Fair by Daniel Burnham, which projected a monumental City Beautiful to replace the core of the industrial city. Burnham's vision of the central city was the prototype for the new civic center and hegemonic Central Business District of the emerging metropolis, and a model for the 1909 Chicago Plan.

Le Corbusier's modernist image also dominated the transformation of central cities in the twentieth century. In the post-war period, the federal urban renewal program helped to complete the skyline that Le Corbusier projected in the Ville Radieuse. Urban renewal was supposed to replace slums with beautiful new buildings. Physical renovation would produce social benefits; conversely, the failure to make these changes would only perpetuate urban problems.

Figure 5.1 Burnham's urban renewal scheme. A portion of Daniel Burnham's rationalistic 1909 Plan for Chicago. Photograph courtesy of The Art Institute of Chicago.

Figure 5.2 1851 plan of Palma Nuova. A planned fortified town near Venice. Credit: Associazione 'ProPalma.'

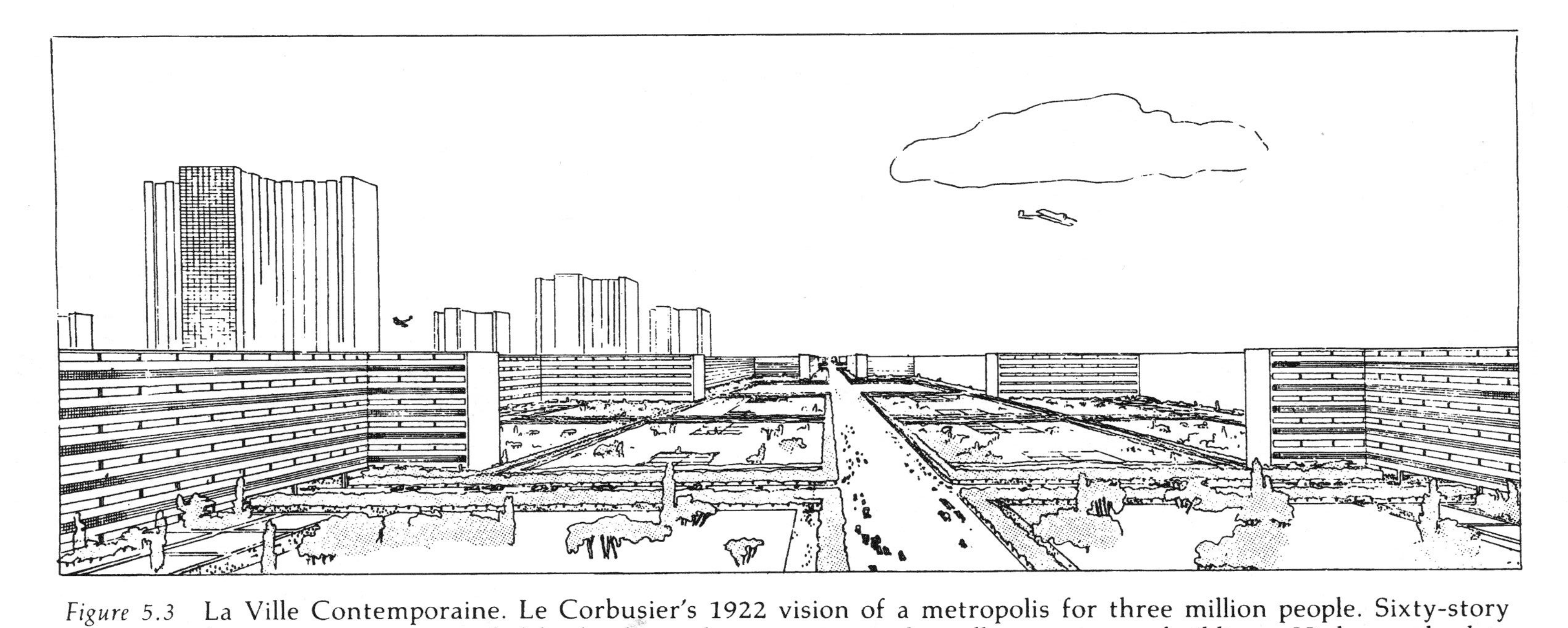

Figure 5.3 La Ville Contemporaine. Le Corbusier's 1922 vision of a metropolis for three million people. Sixty-story towers in the center are surrounded by landscaped open space and smaller apartment buildings. Highways lead to suburbs with single-family homes.

Construction of the new interstate highway system would also make possible the redistribution of industry and dispersal of labor to the suburbs, key elements in the broad modernist vision of the metropolis projected by Frank Lloyd Wright and others.

Physical determinism became a hallmark of the 'progressive' urban reform movement which originated in the early part of the century. This movement was based on a liberal critique of poverty and the slum conditions of the budding metropolis. The movement gave us the first zoning ordinance, in 1916 in New York City. In the following decades, zoning, planning and 'scientific' urban administration by professionals and technocrats were established across the land. Robert Moses, the quintessential reformer-cum-technocrat, fashioned the New York City landscape with monumental parks and highways. He provided the infrastructure for the downtown skyscraper haven and the suburban dream house. To accomplish his grand vision of the metropolis, Moses oversaw the most dramatic displacement of urban population since Haussmann bulldozed the working-class neighborhoods of Paris in the nineteenth century to achieve Napoleon's grand urban vision (Caro, 1974).

An ahistorical approach to urban history often merges with the physical determinist outlook. This approach posits the existence of universal physical structures to which certain eternal characteristics may be attributed, independently of their location with respect to time, place and circumstance. The ahistorical environment is made up of universal symbols or archetypes that define the urban environment from Mohenjo Daro to Los Angeles. Presumably the role of the planner is to identify them and continue their use, even if only as stylistic symbols on modern structures. In the real world, of course, the same physical forms can have entirely different meanings in different historical periods and national contexts.

Gideon Sjoberg's theory of the pre-industrial city takes certain common features of cities in the pre-industrial era and makes them into a universal archetype existing, to one degree or another, in *all* historical periods (Sjoberg, 1960). It matters little to Sjoberg whether we are in the epoch of communalism, feudalism, capitalism or socialism; for him, there is always an ideal-type pre-industrial city. Sjoberg's approach obscures the qualitative difference between, for example, the African village before colonialism and the African village under colonial and neo-colonial domination, a difference often concealed by an outward physical resemblance.

The granary in the pre-capitalist settlements of the Tigris Valley is not the same as the granary in the period of agribusiness and monopoly capital, even if they are physically equal to the finest detail. There never was and never will be an ideal, abstract granary the same for all time.

In medieval Europe, city fortifications served principally as a means of defense; today the same fortifications, where they still exist, are mainly museum pieces, backdrops for recreational uses, or simply obstacles to speculative land development. A narrow, physical determinist approach would fetishize the physical form of the wall and attribute to it some universal characteristic that transcends all time. But the wall in downtown Bologna today has no real meaning to society outside of today's social and physical context.

In a world dominated by international capital, the metropolitan centers of finance capital are sometimes presumed to be universal models for the rest of the world. Western urbanists are particularly fond of ascribing eternal qualities to their skyscrapers or sprawled suburbs.

In sum, the world does not spin on the axis of the architect, the planner or the urban designer.

We have reviewed some of the problems with the anti-urban bias and anti-planning bias in urban theory. We have also examined physical determinism. Even though they may be separate in abstract theoretical terms, in the real world these trends often coexist. They merge to form a conservative approach to planning. For example, the anti-urban bias tells us that immigrants to the metropolis are part of the problem; determinist planners, both private and public, come up with pretty, rational plans to redevelop their neighborhoods; and anti-planning advocates tell us this is all a natural process anyway.

TOWARD A THEORY OF METROPOLITAN PLANNING

Once we get over the anti-urban and anti-planning biases, and reject physical determinism, how do we approach planning for the modern metropolis? The key is in understanding the role of human consciousness.

The thrust of all historical development is toward the fullest expression of human consciousness, as opposed to blind spontaneity, in art, culture, architecture and every aspect of human activity. Planning the social, economic and physical contours of society requires, and assumes, a high level of conscious activity, activity which is made possible as the human race continues its attempts to gain control over nature (though it is also destroying it). Technological and scientific advances, and the production of a surplus of goods and services, set the basis for planning, and all conscious activity, even within the context of free market anarchy.

There is a good deal of planning under capitalism, more planning than in any previous mode of production. There is more planning under monopoly capital than early forms of capitalism. There is more planning in the mature regulated regimes of accumulation than in the relatively unregulated regimes. However, especially in the latter case, planning is always subordinate to the logic of profit and private property. It is utilized by the class of property holders to defend their interests against the propertyless. As in all modes of production, *planning is an extension of economic and political power.*

Conscious planning cannot suppress the objective laws of urban development, or the objective laws of economic development. On the other hand, these objective laws need not enslave the human race. More advanced forms of social development will allow for greater conscious direction, as the human race liberates itself from the laws of nature, private property and need.

Conscious planning is not simply an act of will, but an objective process integrally linked with the material development of society. Objective laws of planning and decisionmaking behavior explain planning and decisionmaking in different social systems. The planning process does not function mechanically, without the conscious activity of the planners, but the conscious activity of the planners is a part of the objective process. Understanding these objective laws is part of being a successful planner.

City planning is the application of human consciousness to the building and preservation of human settlements, and all settlements to one extent or another are planned. Hans Blumenfeld cited the two extremes of the spontaneous squatter settlement and the military camp 'predetermined to the last doorknob.' He correctly rejected the idea that the settlement that evolved spontaneously was unplanned and the predetermined one was planned:

> The difference lies in the degree of consciousness of the city builders. Did they anticipate all their needs and provide for them immediately, or did they have to learn the hard way, being

> forced gradually to adapt the framework of their community to their way of living?
>
> (Blumenfeld, 1971: 3–4)

The real conflict is not between planning and non-planning but over who plans, in whose interests they plan, and whether the plan can stand the test of time (i.e., does it 'work'?).

Professional planners are obviously not the only people endowed with the ability to consciously project and determine the form of the built environment. Selection of a site for the location of a city, or the layout of streets and building lines, is a conscious act, and can be performed by a layperson, a surveyor or a planner with a graduate degree. It may or may not conform to a plan set down on paper by any of the above.

The planning profession is a very recent one historically, and in most places around the world is still undeveloped and dependent on the more narrow field of architecture. The first generations of professionally trained planners have been absorbed into technocratic and bureaucratic hierarchies that limit their abilities to function as planners. Thus, many planners willingly refuse to accept the fundamentally political role of planning, and cede to others the real planning process, the ability to influence and change the world.

Planning for the metropolis requires a qualitatively new approach that takes into account the size, complexity, division of functions, mobility and other particular features of the metropolis. No credible future for the metropolis can be captured in a formalistic sketch or master plan. The territory is too large and the machinery too complex to simply transform the metropolis, or even a part of it, into the realization of an abstract idea from the brain of a master planner. Where this has been attempted it has produced planning nightmares.

The metropolis has produced a highly socialized planning and development process. In the age of the metropolis, consciousness in city building can no longer be that of a single individual. Planning has become institutionalized in the local and national state, in a whole array of institutions that are part of civil society, and often in a community of private landowners and developers. In the more mature metropolitan regions, a systematic process for planning has been established, linking state and civil society, elite groups and technocrats, ruling and minority blocs, in a complex and contradictory struggle for power over decisionmaking. Thus, the metropolis has produced the conditions for the democratization of planning.

With the expansion of the state under all economic regimes, a powerful technocratic and bureaucratic stratum has emerged. City planners are usually part of this stratum at local and national levels. As such, they are not immune from the conflicts within this stratum, or the conflicts between this stratum and other social and political forces. In the metropolis, they are one among many forces, inside and outside government, that determine the form of metropolitan development.

The metropolis does not easily wear the static, linear models from the past because its internal laws produce a rapidly changing division of functions. The metropolis combines all the physical forms that have evolved in over 5,000 years of urban history in a variety of specialized districts with specialized land uses. Therefore, metropolitan planning must be a multi-dimensional and dynamic process. Professional planners, through no fault of their own, are unable to guarantee this dynamism; only a dynamic, democratic political process can guarantee effective metropolitan planning. Planners can and should help understand and interpret the logic of the metropolis and its basic laws in all their diversity. They should know and understand the history of cities and their problems. And they should intervene to help guide the process.

The complexity of the metropolis does not render comprehensive planning meaningless. The answer to the grandiose formality of the eighteenth and nineteenth centuries is not the blind anarchy of anti-planning. This is not an invitation to give up on planning but an appeal to understand the limits of planning. Cities that have evolved with little or no overall planning or control are no paradise. The marvelous medieval cities of Europe and the Middle East, with their quaint narrow streets and intimate neighborhoods built at the human scale, have critical problems overlooked by tourists and heralds of anti-planning such as Tom Wolfe, who never got past walking-tour wisdom (Wolfe, 1981). These neighborhoods lack light and air, and access for emergency and disabled use. Conflict between modes of transport is inevitable because of the narrow streets, and a modern infrastructure is often too costly to install (Venice still discharges raw sewage into the bay). Furthermore, in the Third World metropolis, the inadequacy of an aging infrastructure is inseparable from poverty.

Urban planners in the West seem to have no difficulty rejecting grandiose master-plan ideas, in theory and in practice. However, they usually throw the baby out with the bath water and reject all versions of planning except for those that attribute a special rationality to the real estate market.

Rationality and Planning

Rationality and planning are often equated, so that we are led to believe that planning is by definition a rational activity producing a rational product. This is what is known as an ideological myth. It is a mere hypothesis that has been incorporated in human consciousness as if it were a fact and truth. It is an idea that has become a material force, and that reproduces the social relations which spawn it (see Sorel, 1969).

The greatest sins of planning have been committed in the name of rationality. In practice, rational planning has meant monumentality and allegiance to formalistic and deterministic approaches. Baroque monumentality was an expression of centralized elite power. The 'rational' schemes of Haussmann, Burnham and Speer damaged cities and the people who live in them. Their ideal physical models could work, in a practical sense, in smaller settlements because their simplistic schemes did not have to compromise with the complex and changing needs of the metropolis. They cannot work in the modern metropolis. They will be consumed by the metropolis, like Sunnyside, Greenbelt and L'Enfant's Washington, D.C. They can only be quaint historic districts, or isolated symbols of institutional power like Boston's Christian Science Center, or tiny egotistic post-Modern signatures in the metropolitan skyline, like New York's Battery Park City.

Early concepts of rationality by Smith and Ricardo referred to the rationality of the marketplace. The enlightened monarchies, Bismarck's Prussia and Napoleon all adopted the mantle of rationality to justify their rule. With the expansion of the state under monopoly capitalism in the twentieth century, more comprehensive notions of rationality emerged. In the 1930s, Karl Mannheim gave voice to New Deal style interventionism and projected a global vision of planning for capitalism (Mannheim, 1940). Subsequent theories of the welfare state, such as those of John Maynard Keynes, Gunnar Myrdal and Robert Heilbroner, rely on this more ample notion of rationality.

The broader notions of rationality that emerged in the twentieth century were applied to the modern metropolis, producing the first attempts at comprehensive metro planning. In the West, there were rather feeble and limited planning attempts, because they were subject to the logic of local real estate markets. They included the Garden City and City Beautiful projects. In the East, on the other hand, there were bold efforts to implement comprehensive metropolitan master plans. The rationality of the West concealed the logic of profit and the market, and the rationality of the East concealed the logic of the administrative/command system. The rationality of Western architecture became the rationality of real estate economics. The rationality of Eastern architecture became the serial reproduction of administratively acceptable built forms.

Aside from its use as an ideological myth, and weapon, there is a sense in which rationality has real meaning in planning. It can refer to a systematic, logically coherent *process*. Andreas Faludi (1973) discusses planning theory as essentially a process. Thomas (1982) shows how such a view can obscure the actual content of planning. A rational process has nothing to do with the quality of the planning, or its ultimate historical validity. It can be used to justify the most irrational human acts. It is simply a procedural test. Hitler's architects followed a rational process in planning concentration camps, but there is nothing about concentration camps that imbues them with any substantive rationality.

The federal urban renewal program in the United States was supposed to be an example of a rational planning process. Urban renewal was aimed at eliminating slums and blight through a clearly defined public planning process. Local redevelopment authorities analyzed needs, assessed resources, applied carefully elaborated criteria, and came up with what were said to be logical solutions. Ultimately, this 'rational' process was subservient to special property interests and the redevelopment of central locations by the real estate market. It wrote down developers' costs by managing and financing the acquisition of property, relocating tenants and clearing development sites. It displaced millions of low-income people, a disproportionate number of whom were African-Americans. It was a process whose outcome would be dubbed rational by those who benefited and irrational by those who did not. (See Gans, 1968: 57–77; Angotti, 1981; 1983; Hartman, Keating and LeGates, 1982.)

Planning that follows an orderly process is not necessarily good planning or bad planning, rational planning or irrational. Nor is spontaneous urban development without a formal plan necessarily good planning or bad planning, rational planning or irrational. Again, the question is *who plans*, and *who benefits*?

The most important additions to modern planning theory

emphasize the need to merge a rational, comprehensive planning process with democratic participation. Paul Davidoff was among the first Western planners to link rationality and pluralism in urban planning. In 'A choice theory of planning,' Paul Davidoff and Thomas Reiner (1962) proposed going beyond partial planning and called for public intervention to solve social and economic problems. They believed that establishing a rational process for planning accessible to the public can produce a more just and democratic plan. Davidoff elsewhere emphasized the need for 'Advocacy and pluralism in planning' (Davidoff, 1965).

In a critique of Davidoff, Mazziotti (1982) shows how the concepts of pluralism and advocacy can obscure the real structures of power. However, Mazziotti fails to appreciate the extent to which Davidoff offered a practical theory of action for planners who actually confront power in the political world. As founder of the Suburban Action Institute, Davidoff was instrumental in efforts to challenge the power of exclusionary suburban zoning and achieve integrated metropolitan development.

As suggested by Davidoff, planners should aspire to an ethic of advocacy – for substantively good planning, not simply comprehensive planning in favor of a vague 'public interest,' as suggested by Altshuler (1966), and not simply 'muddling through' as suggested by Lindblom (1959). Planners and planning institutions should seek to engage themselves in long-range comprehensive planning, middle-range planning and short-range decisionmaking. In this, they will be no different than any other political and advocacy group. Their long-range strategies facilitate the ability to influence short-range and middle-range plans. Their ability to influence immediate decisionmaking enhances their capacity for long-range planning. To do all this planners have to acknowledge that they are part of a political process, and have a relationship to political power. As John Forester put it:

> If planners ignore those in power, they assure their own powerlessness. Alternatively, if planners understand how relations of power shape the planning process, they can improve the quality of their analyses and empower citizen and community action.
>
> (Forester, 1989: 27)

The Anglo-Saxon planning tradition which matured in the United States is particularly slanted toward short-range planning. This is related to a regime of accumulation tied to imperial plunder that has historically placed short-term profit ahead of long-range development. The public planning ethic has followed the private planning ethic, and reproduces the inequalities and prejudices it inherits.

Planning must make the leap in human consciousness that corresponds with the qualitative leap from industrial city to metropolis in the twentieth century. As usual, human consciousness is still far behind social reality.

6

METROPOLITAN POLITICS AND PLANNING

What is the city but the people?

Shakespeare, *Coriolanus*

We have shown how the metropolis is a qualitatively distinct form of human settlement that emerged in the twentieth century, a relatively large settlement with a developed internal division of labor including separate functional districts, a central city and suburbs. The metropolis is also characterized by new and diverse forms of planning, governance, management and decisionmaking. It has given rise to new social movements and forms of political struggle. Around the world the metropolis exhibits a new set of political dynamics which were not found in the smaller industrial cities and towns of the past, nor do they characterize the cities and towns of the present.

Understanding these dynamics is a necessary prerequisite to planning. The size and complexity of the metropolis are making the planning process more of a socialized process, and less the work of individual master planners. The rise of the metropolis has increasingly placed planning at the center of the local political process, and made the political process more focused on planning. As a result of the growing division of labor and division of functions, neighborhood and community politics, and neighborhood and community planning, have emerged as significant elements in metropolitan politics. The key to unlocking the potential of metropolitan-wide planning is understanding and working with neighborhoods and communities in the context of a metropolitan vision of balanced and equitable growth. To do this, we must leave behind the antiquated notions of master planning and master building inherited from the nineteenth century.

Metropolitan planning

Metro-wide planning is still in its infancy. Attempts at metropolitan planning around the globe have been limited and fraught with difficulties. They have usually relied on out-dated political and social institutions inherited from the pre-metropolitan era, and planning methods more appropriate to smaller pre-metropolitan settlements. Many have been thwarted by the emergence of competing municipal interests and powerful regional agencies. Municipal governments retain substantial decisionmaking powers that often frustrate metropolitan planning, especially when there are many small municipal jurisdictions. No matter how important municipal planning may be, it is no substitute for metropolitan-wide planning.

One of the fundamental distinctions to be made between earlier city planning and metropolitan planning is that because of the size and complexity of the metropolis, its future can no longer be effectively charted by the old-fashioned static master plans. Metropolitan planning requires a broad vision of the future but not a detailed blueprint. Without a metropolitan perspective the anarchy of local neighborhood self-interest will prevail. However, planning which incorporates neighborhoods is critical to the success of metropolitan planning because neighborhoods are now the fundamental building blocks of cities. This is the main lesson of nearly a century of experiments in metropolitan planning. Peter Hall notes:

> metropolitan planners always fight the last war. . . . In the 1920s and 1930s, they waged battle against the evils of nineteenth-century industrialism just as the market was showing a way out into the suburbs. . . . In the 1950s and 1960s . . . planners sought to impose a steady-state view of the city derived from the stagnant 1930s. In the 1970s, as growth flagged, they produced plans for metropolitan expansion. Only belatedly, during the 1980s, are they coming to terms with a new reality of metropolitan decline.
>
> (Hall, 1984: 431)

The planning lag is more epochal than Hall implies – and what he perceives as metropolitan decline is really the stability of the developed metropolitan areas. The planning lag applies to the entire practice of planning in the twentieth century, in the U.S., Soviet and dependent metropolis. Metropolitan planning is still based on the nineteenth-century notion that the planners, the master builders, are chosen to bring consciousness to city building. This notion has even survived the anti-planning critiques that have all but buried master planning in the West, where the role of conscious master has been transferred from master planners to master developers. Still in its infancy is an approach to metropolitan-wide planning based on the recognition of a strategic role for neighborhoods. This approach is not simply a matter of choice, or political orientation, it is an historical imperative spurred by the rise and evolution of the metropolis. We shall return to neighborhood planning in the next chapter.

A few pioneer efforts at metro-wide planning have had a significant impact on the shape of the metropolis, even where they have only been general and advisory in character. Among the most important are Abercrombie's Greater London Plan of 1944, which included ideas that were later integrated in the post-war New Town program and subsequent strategic plans; the 1948 Finger Plan for Copenhagen; Stockholm's 1952 plan; and the 1965 Schema Directeur for the Paris region (Blumenfeld, 1971: 93–110; Hall, 1984: 435). A metro-wide planning process has also emerged under strong metro governments and regional authorities. These include, for example, the Municipality of Metropolitan Toronto, created in 1953, and Tokyo's National Capital Region, established by the NCR Development Law of 1956 (Hall, 1982; Self, 1982).

In Japan and parts of Europe, which tend to have interventionist urban policies, the national state tends to play a major role in metropolitan development. This is especially true in countries where there are fairly large publicly-owned tracts of urban land. In some countries, such as Holland and Sweden, local governments have extensive powers to condemn developable land in the suburbs and plan future growth (Marcelloni, 1987). In Great Britain and Italy, public acquisition of suburban land was a pre-condition for the planning of new towns and suburbs.

Metro planning experiments throughout the world have had a somewhat difficult time, and many have been abandoned or substantially scaled down. In Chapter 4 we noted both the accomplishments and problems of the Moscow Master Plan, and how early attempts to limit the size of the metropolis were abandoned.

Planning for metropolitan Tokyo offers further lessons on the limitations of planning in regions with high rates of growth. The 1956 NCR Development Law delineated an urbanized area about ten miles from the center of Tokyo beyond which a seven- to ten-mile-wide green belt was to be established. Beyond the green belt,

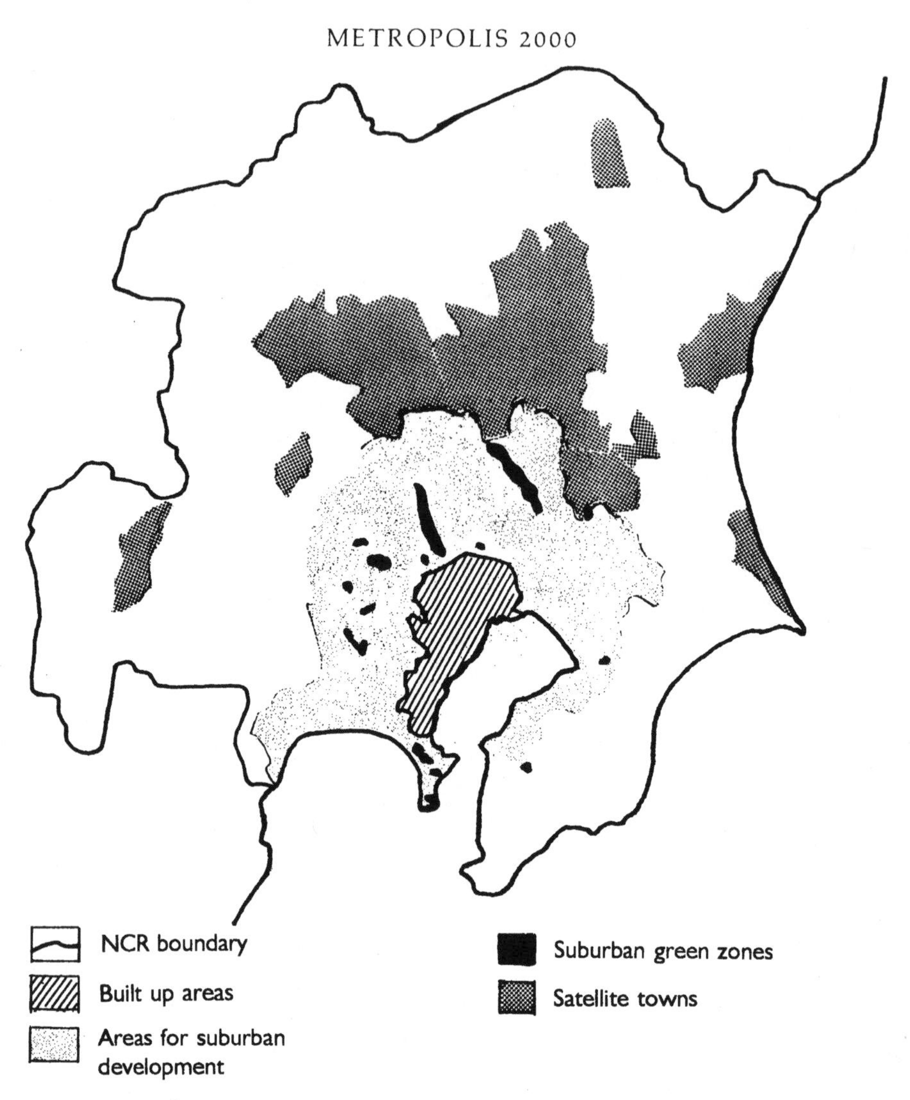

Figure 6.1 Tokyo national capital region plan, adopted in 1986.

planned satellite communities would emerge to absorb new demands for growth. The approach was similar to Abercrombie's 1944 plan for London, and had somewhat similar results. Tokyo's metro plan was backed up by national legislation and could rely on a relatively powerful national executive with central control over regional infrastructure. Municipalities, including the central city Tokyo Metropolitan Government, were subject to the plan. While the plan produced somewhat orderly growth in some areas, however, it failed to provide adequate mechanisms for controlling the extent of growth. Limits on development in central Tokyo gave way to land market pressures. Plans for the green belt were swamped in a flood of new development. As a result, the NCR Development Law was revised in 1965 to incorporate new growth within the green belt (Hall, 1966; 1984).

A more successful metropolitan planning process may be found in Sweden, where a relatively high standard of living, national welfare state, cooperative approach to local government, and moderate rate of urban growth converge to provide the conditions for sound planning. Metro planning in Sweden is mandated and supported by national legislation that requires that municipal governments in metropolitan regions join together and develop a regional plan. Since the beginning of this century, municipalities have had powers to condemn undeveloped land at the urban periphery. Municipalities are required to have master plans, which have the force of law, and these must be consistent with the regional plans. Zoning regulations and building regulations follow from the guidelines established in the master plans. While central government authorities may have little direct involvement in plan preparation, they can withhold financial assistance to municipalities that attempt actions not in accordance with the plans (Walsh, 1969: 195, 206).

The Stockholm metropolitan region includes the central city and scores of planned satellite suburbs strung along suburban rail lines. Its total population is 1.5 million. The central city includes a preserved historic center and a modern business district that resulted from a controversial redevelopment plan that displaced many people in the 1960s. The first suburban development began in the 1940s on agricultural land that had been acquired by the City of Stockholm in the preceding decades. The density of the suburban centers is similar to the central city density, and neighborhoods are oriented around high-rise centers and superblocks. The master plan for Stockholm, first elaborated in 1952 and subsequently revised, is developed and implemented by a regional association of local governments including the City of Stockholm and twenty-two other jurisdictions. Its implementation involves the Greater Stockholm Building Board (a voluntary inter-municipal association), housing cooperatives, national government agencies, city, county and municipal governments.

It is useful to note that the planned development of Stockholm occurred without a single powerful metropolitan government. More important than metro government, perhaps, is single ownership of land, and the wide consensus that emerged around a single metropolitan plan and a unified process of planning. While there were conflicts among local governments, there was also a political culture which relied on cooperative decisionmaking (Walsh, 1969: 195, 207; Odmann and Dahlberg, 1970).

In the United States, metropolitan growth has taken place largely within the context of inter-municipal competition and private real estate markets shaped by national infrastructure subsidies and policy. Municipal governments, of which there are many within metropolitan areas, often develop plans, but their master plans have two crippling deficiencies: they are no less susceptible to the grand master-plan fallacy than metro-wide planning, and they are often not coordinated with any metropolitan-level planning and decisionmaking. The U.S. is the most metropolitanized large country of the world, yet has had the least experience with metropolitan-level planning (Perloff, 1980). We shall discuss reasons for this later on.

Aside from the experiments in metropolitan-level planning, there are many examples of inter-municipal cooperative efforts (see Bollens and Schmandt, 1970: 349–72). In the United States, county governments and regional planning agencies in some states formulate metropolitan development strategies. However, these strategies are often compilations from local (municipal) plans, reduced to their lowest common denominator and shorn of conflicting elements. They are ideas with no operational government body to implement them or to regulate private development. As a result, planning practice tends to follow the spontaneous patterns of urban growth. Unlike the previously mentioned metro experiments, there is often no conscious, publicly acknowledged vision of a future metropolis.

In centrally planned economies, there has been a wealth of experience in metropolitan-level planning, beginning with the pioneering 1935 Moscow Master Plan. At least until 1989, most Soviet, East European and Chinese cities had operative comprehensive plans, and metropolitan- or national-level governments responsible for implementing them. They were not always implemented as originally laid out, but they were usually more than reports that stayed on shelves.

The experiences in socialist countries show us both the possibilities and limitations of metropolitan-wide planning. One of the problems has been an over-reliance on a single, rigid master plan. Another has been that, despite the formal acknowledgment of a basic neighborhood level for planning – in the USSR, the *rayon* of about 100,000 population – a neighborhood-level planning process has often been absent. Indeed, neighborhoods have been considered homogeneous products of a rigid system of city building, modules to be reproduced by applying assembly-line production principles (see Chapter 4 for further discussion).

There are also many examples of metropolitan planning in the dependent metropolis of the South – perhaps more than in the metropolis of the North. The experiences of India and Brazil are significant in this respect. However, the problem in the Third World is usually a lack of human and material resources for implementation. Even where there are strong national and metropolitan governments – a legacy of colonialism – the ability to build and regulate building according to master plans is limited. The ability to plan with neighborhoods is even more limited. It is no surprise, therefore, that the implementation record of metro planning in the Third World is fairly poor.

Calcutta is a metropolis of almost 13 million, the largest in India. Since metropolitan planning was instituted in the years after independence, it has been unable to keep up with the dramatic in-migration from rural areas and Bangladesh. Before independence, the 1909 Town Planning Act and 1911 Calcutta Improvement Act established administrative mechanisms for guiding urban growth. The Calcutta Improvement Trust served as a sort of comprehensive development corporation to plan and carry out public works and rationalize urban development. The State of West Bengal, of which Calcutta is part, has had broad powers to undertake comprehensive planning, and could mandate planning in the thirty-five municipal bodies in the Calcutta metropolitan area. These mechanisms, however, proved incapable of dealing with the rapid rate of urban growth, extensive poverty and inadequate resources.

In 1961, the Calcutta Metropolitan Planning Organization (CMPO) was created within the Calcutta Metropolitan District, and in 1965 the Calcutta Metropolitan Planning Control Act was passed. The CMPO produced a *Basic Development Plan* (1966) that outlined a broad perspective for the future growth of the region and a five-year development program. In 1970 the Calcutta Metropolitan Development Authority was created to coordinate new development. In 1976, India passed the Urban Land Ceiling and Regulation Act, which was to stop land speculation and socialize land use by placing a ceiling on the price of vacant land (Bose, 1980).

At the time of the 1966 master plan, Calcutta's population was about 7.5 million. In less than twenty-five years, it nearly doubled. While some new districts were comprehensively planned, Calcutta mostly grew spontaneously and without government planning. The master plan's goal was to improve life for the million or so living in shantytowns, and establish 'community facility cores' there; in the long run the shanties were to be cleared. Instead the population in shantytowns expanded and was joined by the legions of pavement dwellers. The plan was to set up industrial growth poles in Calcutta's periphery that would serve to decongest the central areas. In the context of India's national economic plan, this strategy could have been implemented. However, India's growth poles mostly ended up outside metropolitan areas and fared no better than those in Italy, Venezuela or South Africa (see Chapter 3). According to one scholar, most development plans in India were little more than advisory and 'few of them have achieved more than token implementation' (Sundaram, 1977).

Urban planning in India followed the tradition of British idealism, both before and after independence. This tradition rests on formalistic anti-urban notions of the ideal urban environment, and complex administrative mechanisms to achieve it. In developing nations with limited resources, the potential for realizing these visions is minimal. The CMPO was created by the State of West Bengal and its plan was drafted with international technical and financial assistance. There was minimal participation and erratic support from the thirty-one municipalities within the Calcutta Metropolitan District. In a metropolitan area that effectively evolved as a conurbation around company towns (many of them founded on the banks of the Hooghly River at the site of jute mills), such oversight was fatal to the plan (see Walsh, 1969: 196; Sundaram, 1977).

The universal experience of metropolitan planning so far is that formalistic master plans are not worth the paper they waste to have them printed. This much is obvious to planners in every part of the world, even where some of these plans have been implemented, or perhaps *especially* where they have been implemented. The real question now is why? Is it because master plans themselves are useless, or is it the content of the plans? Should planning be reduced to a process alone, as often asserted by those who believe planning can never play a substantive role in urban development?

The first condition for metropolitan planning is projection of a conscious vision of preservation and development of the region. This can be done through a master plan proposing new growth areas, an unofficial document put together by technocrats, or an extended process of community debate and decisionmaking. In the New York metropolitan area, for example, the private non-profit Regional Plan

Association, founded in 1922, has done more to promote a metropolitan-wide vision than any governmental body. This vision reflects the collective interests of a corporate establishment, though sometimes at odds with individual property interests (see Fitch, 1976).

The second condition is that there must be a unified economic and political mechanism to implement that vision. If the plan proposes to give shape to new growth areas, it will not be implemented unless the main parcels of vacant land are under *single ownership*, and there is a unified political jurisdiction. Fragmented ownership and government cedes power to the local real estate market and local elites; it thwarts unified planning. This does not mean that there can be no planning under small-scale private property or fragmented local government, only that the planning will be subject to the self-interest of parochial property and political factions.

Urban infrastructure is the most strategic element in metropolitan growth and perhaps the most amenable to planning efforts. Indeed, unless metro-wide government and planning are integrated with the maintenance and development of the physical and social infrastructure, there is little hope for implementation. The specialized needs for infrastructure development have given rise to specialized sectoral agencies and independent authorities responsible for planning roads, water, sewers, utilities and other infrastructure. Since the basic infrastructure is usually regional in character and requires major capital expenditures, efficiency and scale economies prompt metropolitan-wide approaches even in a market economy with strong municipalities (see Stein, 1988). In the U.S., independent authorities flourish in states with constitutional limits on municipal lending. Water, sewer, transportation and other authorities can seek independent bond financing for the special purpose(s) they are mandated to perform.

Sectoral agencies and authorities often develop independently of elected government and may resist participation in metropolitan-wide planning. They tend to be protected by the technocratic mystique and special legislation, and often become narrowly focused on maximizing their income and number of employees, or expanding their jurisdiction. Since independent authorities usually resort to bond financing, they are compelled to set user fees for their services to guarantee investor profit and insure growth, without necessarily structuring service levels and charges to meet land-use or social objectives or to achieve equity. For planning that is more comprehensive than sectoral planning, the intervention of a higher central authority – either metropolitan or national government – is needed.

There are also multi-purpose districts, such as the Port Authority of New York and New Jersey (PANYNJ) in the U.S., perhaps the largest in the world. The PANYNJ is a bi-state financial empire that owns and operates five marine terminals, three international airports, four toll bridges, two tunnels, three bus terminals, three industrial parks, a rail line, the World Trade Center and other facilities. The Authority has an annual budget of $1.4 billion and its facilities and allied businesses employ over 425,000 people. It is, in effect, the most powerful planning agency in the New York region. As with other multi-purpose districts, it can readily become an independent bureaucracy and immune from cooperative planning efforts or public oversight.

Government in the metropolis

The experiences in metro planning suggest that metropolitan government can provide a critical institutional framework for successful metropolitan planning, even if it is not an essential element. Yet metropolitan government, where it exists, is either weak and undeveloped or so overcentralized it cannot cope with the real problems of land use and planning (see Walsh, 1969; Horan and Taylor, 1977).

Most national constitutions and government policies treat the metropolis as no different than any other form of settlement. The most notable exceptions are federal districts or capital city regions like Washington, D.C., Brasilia, and Mexico, D.F., which have unique juridical statuses, sometimes favorable and sometimes unfavorable to democratic governance. In the case of Washington, D.C., for example, its unique legal status means lack of democratic control for the mostly African-American central city population. Another problem is that federal district boundaries tend to be inflexible and often do not change to coincide with the limits of the developed metropolitan area. For example, the Mexico City and Washington metropolitan regions now extend far beyond the federal districts.

Where metropolitan government does not exist, there is a potential for building it from the bottom up. For example, the United States has practically no experience with metro-level government, yet its high level of metropolitanization has created unprecedented opportunities for its establishment. The U.S. has a political tradition that would most lend itself to a progressive and flexible devolution of power from state and national government to the metropolis, and at the same time incorporate local communities and neighborhoods in the process. The U.S. Constitution gives broad residual powers to the states, which in turn have delegated substantial powers to municipalities. The U.S. does not have the feudal legacy of Japan, the Bonapartist tradition of Europe, or Great Britain's imperial bureaucracy. The ideology of local control and opposition to central government is deeply imbedded in U.S. political culture.

But local control in the U.S. is part myth and part reality. Despite a tradition of local revenue generation through real estate, sales and income taxes, local governments have proven incapable of generating sufficient resources to address the problems of poverty, drugs and crime, to clean up pollution, or undertake large-scale public works such as mass transit or highways. The municipalities in large metropolitan areas, especially central city municipalities, rely on federal and state governments for greater proportions of their revenues than cities and towns, and even these amounts are insufficient to meet urgent needs. This remains true even after the Reagan years of federal cutbacks to local governments.

The U.S. does not have a metropolitan form of government, or a clear national policy focusing on metros, even though it is the most highly metropolitanized nation in the world. The closest thing to a metropolitan government is the Dade County Metro (Sofen, 1966). Los Angeles County encompasses most of the Los Angeles region, but its powers are more limited. In any case, these are the exceptions to the rule. County governments all over are weak, and mostly mediate the competing interests of the many municipal governments within their jurisdictions. There are also inter-municipal councils of government (COGs), inter-local agreements and multi-purpose regional districts. In many ways, however, their role is more advisory than directive or operative (see Bollens and Schmandt, 1970: 364–8).

The U.S. has perhaps the most fragmented system of local government (Coulter, 1968; Alexander, 1982; Young and Garside, 1982; Pinch, 1985). The average U.S. metropolis has over 300 local governments (Bollens and Schmandt, 1970: 102). This fragmentation facilitates an exclusionary approach to governance and planning based on the hegemony of conservative suburban elites (see Chapter 2).

For a brief period of time, from 1964 to 1979, the federal government in the U.S. had something of a national urban policy giving priority attention to the elimination of urban inequalities. In the 1960s, President Lyndon Johnson's War on Poverty brought federal funds directly to large central city governments, bypassing more conservative state governments. At the same time, however, federal funding of suburban development was used to insure uniform compliance with the standards of the sprawled, auto-based suburban pattern promoted by the auto-petroleum and national pro-growth blocs. The federal commitment to bypass conservative state governments was reversed during the Nixon and Ford Administrations (1969–76), and finally buried by President Ronald Reagan's 'new federalism,' which cut federal funds to local governments under the guise of enhancing local control (see Gartner, Greer and Riessman, 1982). National urban policy dissolved into a new version of the conservative 'state's rights' doctrine, federal programs continued to favor the affluent suburbs, and central city needs were once again held hostage to conservative state legislatures and governors. As central city governments were increasingly led by African-American mayors, their control over financial resources dwindled.

Most municipal boundaries in the U.S. were drawn in previous centuries, when the nation was primarily agrarian. The trend in the twentieth century has been for new suburban municipalities to incorporate independently of the central cities and other suburbs. After a wave of annexations that ended around the turn of the century, few local governments have since joined together so that their boundaries could conform with the developed metropolitan areas. There have been some city–county consolidations and annexations, but these have not fundamentally altered the structure of local government (see Bollens and Schmandt, 1970: 297–307).

Most of the large metropolitan regions in the U.S. have never had even general long-range perspective plans. The New York metropolitan region exemplifies the chaotic proliferation of uncoordinated local governments. The region has over 2,000 units of government including three states, thirty-one counties, 780 municipalities, 716 special districts and 661 school districts (Danielson and Doig, 1982: 4; Institut, 1988). The only metropolitan-wide planning agency is the private non-profit advisory group, the Regional Planning Association. The RPA has performed useful studies and regional plans but has no formal decisionmaking role. Short-lived federally-funded regional transportation agencies have dissolved. There has never been an official master plan for the New York metropolitan region. A draft master plan for New York City, the core municipality, was completed in 1971 but never adopted.

It has long been known that the New York metropolitan region pays out more in taxes to federal and state governments than it gets back in revenues (Carey, 1981). At the same time, there is a major gap between the revenue needs of the central city and suburbs. The central city tends to have a larger proportion of people living below the poverty line, and greater needs for housing and social services, yet provides more services for the region as a whole than most of the suburbs. The central city, with state and federal assistance, spends more per capita for social services, but less for education and other services related to the quality of residential neighborhoods. These are the consequences of local government fragmentation.

There are over 75,000 local governments in Europe. Their powers range from very weak to very strong depending on the national context. France, which has one of the strongest national governments, has the largest number of local governments – over 36,000 – and few metropolitan governments (Sharpe, 1979; B.C. Smith, 1985; Borja, 1988).

Some European countries have metro-level governments, though their powers vary considerably. Some of the governing bodies are directly elected and others are indirectly elected by local municipalities. Functions range from provision of basic services to planning and coordination. Among the more active metropolitan governments are: the Metropolitan Corporation of Barcelona, which coordinates assistance to twenty-seven municipalities (but which has recently been weakened by political changes); the Community Council of Madrid; the Stockholm County Council; the Agglomeration Council of Brussels; the Ile-de-France Regional Council; and the Comune of Rome.

In 1990, Italy enacted a unique comprehensive national law that is expected to radically restructure local government by creating between nine and thirteen metropolitan governments. From one-fourth to one-third of the nation's total population will fall within the orbit of the newly-defined governments. They will have broad powers now vested in the regional, provincial and municipal governments, including planning, transportation, waste management, water management and environmental protection (Sistema Permanente di Servizi, 1991).

In Great Britain, the Greater London Council was a model for regional coordination and planning. Its rise in the 1970s represented

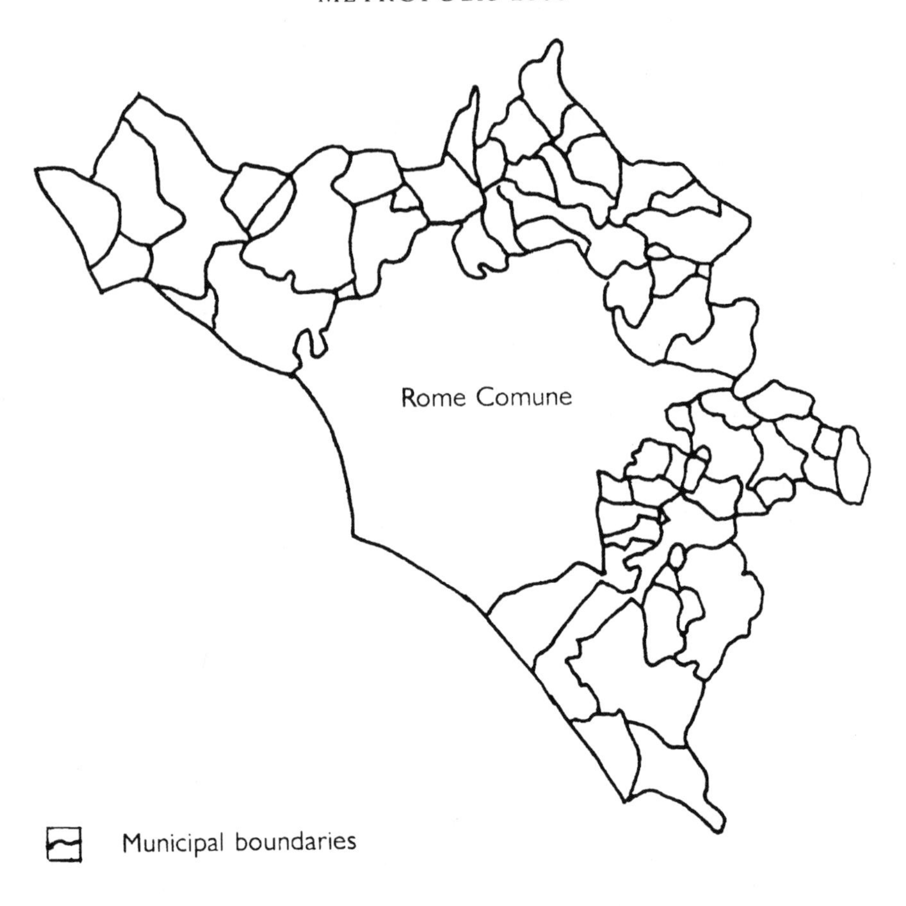

Figure 6.2 Rome metropolitan region. Under Italy's new law creating metropolitan governments, the Rome Comune, which already encompasses a large portion of the built-up metropolitan area, will join with at least sixty-four other municipalities.

the merger of inner-ring suburban with central city interests (see Alexander, 1982; Young and Garside, 1982). In 1986, the Council was disbanded under the austerity measures of the Thatcher government. In Canada, there is the Municipality of Metropolitan Toronto and the Montreal Urban Community. In Japan there is the Tokyo Metropolitan Government, under the elected Metropolitan Assembly, which serves twenty-three special wards, twenty-six cities, seven towns and eight villages (Institut, 1988).

With the restructuring of capital and development of a new international division of labor, there may well be a new thrust in Europe and other developed regions toward stronger local government, as Magnaghi (1982) has suggested. Decentralized production may well encourage decentralized urban decisionmaking. This may also coincide with the resurgence of regionalism and ethnic politics in Europe (Hadjimichalis, 1987; Rudolph and Thompson, 1989). However, these trends are in opposition to the dominant processes of European integration, centralization of decisionmaking and metropolitan hegemony. Gottdeiner (1987), for one, speaks of the death of local politics. Perhaps the European unification process will produce a more uniform system of local and metro government in the next century, but it is not clear whether local authorities will gain or lose in power as the individual national governments seek to protect the powers left to them.

In the Third World, metropolitan-level governments are widespread. They include, for example: the Brazilian metropolitan areas, where there are inter-municipal Advisory Councils and appointed Deliberative Councils in all large cities; India's metropolitan districts; the Metropolitan Municipality of Greater Istanbul; the Jakarta Metropolitan City; the Karachi Metropolitan Corporation; the Municipality of Metropolitan Lima; and the Metropolitan Manila Commission (Institut, 1988).

Despite their prevalence in dependent countries, metropolitan government, and all local government, tend to be creatures of national political power, more so than in Europe and the United States. This is due in part to the colonial legacy, which has resulted in more centralized national governments and a minimal role for municipal government. Colonial rule required strong central control and weak local authorities. To the extent that local authority was allowed to develop, it was located in the colony's capital city, which invariably became, later, the independent nation's major metropolis. In developing nations, we often find a highly developed formal system of governance in the metropolis contrasted with a poorly developed system in other cities and towns. Political power in the metropolis is usually only one step away from national power. But everything outside the metropolis is seen as somewhat provincial and politically inconsequential.

In the socialist countries, the problem of fragmentation and obsolete municipal boundaries has been less significant. All metropolitan areas tend to be governed by metropolitan-level governments. Still, these governments, and local government in general, have been relatively undeveloped. In the Soviet Union and Eastern Europe, the administrative/command system became increasingly overcentralized even as the need for decentralized government increased. Furthermore, concentration of power in the Communist parties thwarted development of independent local authorities. The Soviet legal system theoretically permitted a wide latitude for metropolitan government through local Soviets. However, this potential was not realized, as recognized during the period of *perestroika* when a process of strengthening local government was initiated. With the demise of the Soviet Union and independence of its constituent republics, it remains to be seen to what extent new powers will accrue to metropolitan as opposed to republican governments.

The main problem everywhere remains one of establishing metropolitan-wide government with appropriate powers. This is especially true in the United States, which is the most metropolitanized country but has one of the most fragmented systems of municipal government. If there is no metro-wide planning context, municipal governance can easily mean government of exclusion. If there is no metro-wide planning to equalize differences between central city and suburbs, municipal governance can institutionalize and legitimize inequalities.

The problems of metropolitan governance stem in part from the fact that we live in an epoch in which nations are still the dominant form of political organization and in which national states dominate local government. Power and resources are centralized in national governments, and the local state is almost universally subject to national control. Dominant political forces at the national level are not necessarily threatened by opposition control over local governing bodies. It is perfectly possible for some political parties to control many or even all local governments without having access to the main levers of power in a country. Witness the case of the Italian Communist Party (now the Democratic Party of the Left), which for forty-five years had been the majority party in northern Italian municipalities, but continues to have its reforms frustrated by Christian Democratic control of national government.

As the metropolis and a new international economic order evolve, the old system of national–local governance will probably become more and more obsolete. The need for a new type of social formation, something on the order of a metropolis-state, may grow as international political and economic integration proceeds apace, and as national governments find it more difficult to control government at the local level.

Metropolitan politics, municipal reform and pro-growth blocs

It is difficult for many professionally trained planners to accept that planning is a fundamentally political process. Metropolitan planning grows out of and reflects the dynamics of metropolitan politics, the aspirations and protestations of professionals notwithstanding.

Metropolitan politics is in turn shaped by property relations – both urban property and property in the main means of production. However, it would be a mistake to view all politics as direct reflections of urban property or economic relations. There are many mediating factors, including national politics, culture, history, struggles between social strata and institutions, and the role of bureaucracy. The metropolis itself produces and reproduces new social relations that intersect with, influence and mediate relations of production and consumption, and relations between and among classes and social strata. Thus, in his analysis of urban development in the United States, Michael Peter Smith rejects mechanical economic determinism:

> It is the historically specific social and political processes through which economic forces must work rather than general laws of capitalist economic development which best account for the uneven pattern of urban development in the United States. This is why the particular forms of uneven development – sprawling suburban development, widespread urban fiscal stress, extreme class segregation of residential communities, and pronounced population deconcentration of affluent communities to the hinterlands – are not found in this same form in other advanced capitalist states.
>
> (M.P. Smith, 1988: 6)

National governments mediate and interpret the needs of capital and are the key institutions for making urban policy. In some cases, the captains of finance exercise considerable direct influence in urban politics, and even volunteer their time and know-how to help make cities run. Nelson Rockefeller's personal involvement in New York City during several decades testifies to the stake of capital in the politics of strategic cities. However, this direct involvement is the exception rather than the rule. Rockefeller used his New York base to influence national policy, which is of greater strategic interest to capital.

In countries with relatively unrestrained real estate markets like the

United States, small and medium-sized urban property interests and the local state play a crucial role. In the largest financial centers, there is a substantial overlap between property interests in the urban core and international finance capital. However, though the two are natural allies, there are substantial conflicts between them.

Metropolitan politics revolves more and more around issues that have traditionally been considered 'consumption' issues (on metro politics, see Greer, 1962; Coulter, 1968). The industrial cities of the nineteenth century erupted with struggles over wages and working conditions. The metropolis in the twentieth century bristles with struggles over neighborhood displacement and the quality of the urban environment. These are as much economic issues as wages, because they ultimately affect production, including wage levels and working conditions. They are also class issues, because different classes, and different social strata, engage in the struggle to control the process of consumption. However, economic and class issues are increasingly taking the form of struggles over the location and distribution of the social surplus, i.e., how much of a surplus should go to the metropolitan infrastructure, whether it should go to national- or local-level governments, how much of it should be restored to capital via user fees and debt repayment, and which neighborhoods, classes and strata should enjoy its benefits. They are struggles, in short, over the form and substance of social reproduction (Cockburn, 1982).

Metropolitan politics is also shaped by cultural values and symbols that set the boundaries of local decisionmaking. This explains why in so many places a high value is placed on historic preservation of individual monuments and central urban areas, despite the potential market value for redevelopment of these areas. Walter Firey (1945) made this point in his classic explanation for the preservation of the historic Beacon Hill neighborhood in Boston, which is located on some of the most desirable land for high-rise development. However, urban places are not just important cultural values for the Beacon Hill elite. They are also important for the South End working-class community (Gans, 1962), Africans in Soweto, squatters in Lima and many neighborhoods threatened by displacement.

If we trace the evolution of metropolitan politics and planning in the U.S. during the twentieth century, we find it to be bound up with a pro-growth urban reform movement, a political movement wedded to the process of metropolitan expansion (Mollenkopf, 1983). Urban reform became synonymous with building efficient local government through professional planning and management expertise. It has also become synonymous with growth: planning and regulation are to facilitate and promote growth, never to hinder it. Preservation of buildings and neighborhoods is not the main priority, and is seen as useful only insofar as it provides a favorable environment for growth. Logan and Molotch talk about the perception among elites of the metropolis as a 'growth machine':

> The desire for growth creates consensus among a wide range of elite groups, no matter how split they might be on other issues. . . . Although they may differ on which particular strategy will best succeed, elites use their growth consensus to eliminate any alternative vision of the purpose of local government or the meaning of community.
>
> (Logan and Molotch, 1987: 50–1)

In the United States, a young nation where the previously existing urban institutions were only loosely entrenched, the 'municipal reform' movement took immediate hold in the early part of the century and has been virtually synonymous with the rise of the metropolis. It evoked capital's most concerted and conscious approach to metropolitan development, and has garnered a surprisingly high level of legitimacy.

On the national level, the Wilson Progressive era had brought to the fore the idea of rational planning and management through a more interventionist state (see Weinstein, 1968). The underlying premise was that cities, especially central cities, could not be planned or developed rationally through the interplay of competing local politicians, each representing their own territorial interests and none organically committed to the public interest. Under the new 'scientific management,' the Central Business District was strengthened as the leading 'community' in the metropolis, and outward expansion in the suburbs was promoted through major public works. Within this framework, central city governments were to have centralized structures while the new suburban neighborhoods would remain fragmented into many municipalities, each preoccupied with the protection of their own local property interests. The formula was centralized management, rational planning and reforms in the central cities and decentralization, *laissez-faire* and fragmentation in the suburbs. In the guise of serving some vague 'public interest,' it strengthened CBD control in central cities and the rule of suburban elites.

To accomplish their objectives in central cities, the reform movement would have to dislodge working-class, ward-based politics and replace them with city-wide, technocratic politics. They would have to divert local government from protecting and subsidizing working-class territory and concentrate on fortifying business territory. In the suburbs, they would have to guarantee a free hand for growth, either through metropolitan government or the traditional *laissez-faire* system of municipal fragmentation (the latter alternative quickly won out).

The urban reform movement opposed city administrations rooted in the immigrant communities of the Northeast: for example, Tammany Hall in New York and the Fitzgerald administration in Boston. The reform program included promotion of a short list of at-large elected officials instead of the ward-based officials responsive to ethnic neighborhoods, many of them working class in origin. The program called for independent, professional public administrators, comprehensive planning, zoning and rational budgeting.

The new movement followed on the work of earlier reformers like Jacob Riis and Lawrence Vieller, who exposed the miserable conditions in immigrant communities and tenement houses at the end of the nineteenth century. The early reformers had generated the pioneering 1901 tenement house legislation in New York that sought to eliminate slum conditions. In effect, the late nineteenth century reformers helped establish an interventionist role for government in central city areas (see Boyer, 1983).

A hallmark of the new interventionism was the introduction of city planning. As mentioned previously, Daniel Burnham's grandiose image of the renewed central city at the 1893 Chicago World's Fair was partially reflected in the 1909 Chicago Plan. It would later give rise to the mammoth urban renewal schemes that gutted every major city in the nation. New York City was the first metropolis in the U.S. to pass a zoning ordinance that was to introduce rational order into the chaos of city building, which was previously directed by subdividers and surveyors. The 1916 zoning law set the pattern for the rest of the century by speaking to the concerns of downtown business. The law created an exclusive zone for the Fifth Avenue shopping district that would prevent the nearby garment industries, filled with immigrant workers, from further encroaching on their elegant domain. It also established rules for the emerging skyscraper boom. Subsequently, zoning became the mechanism for suburban municipalities across the country to establish exclusive residential areas.

It was no accident that the municipal reform movement coincided with a wave of anti-immigrant hysteria. Capital at the time was concerned about the rise of the labor and socialist movements. The International Workers of the World (Wobblies) were growing, the Socialist Party candidate for President, Eugene V. Debs, won almost a million votes in 1920 even though he was imprisoned during the campaign, and the Bolshevik revolution in 1917 had inspired immigrant workers around the world to attack their own bourgeois institutions. In this environment, 'cleaning up' governments dominated by immigrant neighborhoods took on a special meaning. The national hysteria brought about the herding and deportation of thousands of immigrants by Attorney General Palmer and produced the circumstances that allowed a Massachusetts court to condemn to death two innocent working-class Italian activists, Nicola Sacco and Bartolomeo Vanzetti, on bank robbery charges. (A proclamation by Massachusetts Governor Michael Dukakis in 1977 declared that the trial and appeal were 'permeated by prejudice' against immigrants.)

Municipal reform was infinitely more genteel and rational than the police. It was unmistakeably a WASP movement, though there was plenty of room in its egalitarian framework for the sons and daughters of immigrants. Robert Moses in New York City was an example of this ecumenism. He was an engineer, a professional educated at Yale. Though of Jewish origin, he sought to place himself above ethnic politics; though a Democrat, he sought (and did) place himself above Tammany Hall. Though a city and state official, his perspective was regional; his public works projects, including major highways and parks, had a far-reaching impact on the nation's largest metropolitan area (see Caro, 1974). Moses was an urban reformer with a metropolitan vision consistent with the hegemonic pro-growth philosophy.

A contemporary of Moses, Fiorello La Guardia, perhaps the most colorful and dynamic New York City mayor of this century, was also a creature of the urban reform movement. A fusion candidate, he fought the political machine to win a city-wide consensus encompassing Democrats and Republicans, labor and capital. La Guardia's 'common sense' approach to city problems emphasized the municipal reform movement's attempt to solve problems by cutting through patronage, special interests and encrusted bureaucracy. He fought alternatively the regular Democratic Party machinery, big business and labor, but in the depths of the Great Depression he was the foremost advocate of government as the facilitator of growth. This was the basis for his alliance with Robert Moses and other technocrats. He also brought sections of organized labor into his political coalition to become the foremost urban populist candidate with a corporativist program (see Elliott, 1983).

The availability of a large social surplus after the end of the Depression and World War II fed the pro-growth coalitions in every metropolitan area, particularly in the expanding suburbs. The coalitions received their blessings from the auto and petroleum monopolies, which dominated the national economy (in the 1960s, six of the top ten corporations were in this monopoly bloc). The other main pillars of support were rapidly growing finance and real estate sectors, the construction industry, and the federal government. The coalitions favored central city urban renewal and suburban expansion and were fueled by national legislation favoring metropolitan growth: the federal urban renewal program, the interstate highway program, the Veterans Administration (VA) and Federal Housing Administration (FHA) mortgage finance programs, federal water and sewer programs, and so forth (Mollenkopf, 1983).

Pro-growth coalitions are not simply bourgeois clubs. They make up political *blocs* which include fractions of labor, civic organizations, small and medium-sized entrepreneurs, community and neighborhood groups, technocrats and intellectuals. At times these blocs have gone against the bare economic interests of the monopolies, but for the most part they respect economic power. As with all political blocs, they constantly shift as different forces obtain relative hegemony and others are forced into relative subordination. They also differ in composition from one metropolis to another. In short, we may well speak of many blocs and not just one. In the U.S., capital tends to be hegemonic; labor has played a reduced role, especially since the McCarthy era, and the resources available for urban growth, including surplus capital, land, building materials and energy, have been relatively plentiful, thereby strengthening the pro-growth mentality.

The days of the pro-growth blocs are not yet over but the pro-growth pattern has been seriously called into question. In the United States, one symptom of this is that there is practically no new construction of interstate highway links, especially in central cities. It took the federal government twenty-five years before it finally gave up on the ruinous Lower Manhattan Expressway and Westway plans in New York. The federal urban renewal program and other categorical assistance to local development projects is a thing of the past. In New York City, neighborhood opposition to capital's newest megaprojects has increased the price tag so as to exclude all but the most patient, and wealthy, developers. Even the polite silk-stocking constituencies in Manhattan's Upper East Side, arguably the wealthiest neighborhood in the world, and the quite proper Municipal Arts Society, have launched the charge of 'overdevelopment' in Manhattan.

The booming sprawled metropolises of the Southwest, made to order by the auto–petroleum monopolies for endless outward sprawl, are also evoking cries for an end to development. In the last decade or two, growth in places like Los Angeles and Houston has focused on the downtown cores and existing commmercial centers as well as sprawl at the periphery.

It would be premature to announce the final demise of the pro-growth era. Rather, what is happening is that the old model of growth has become too economically and politically costly, and the most forward-looking investors in urban property are calling for a new model. In part this reflects capitalist restructuring, the proliferation of much smaller units of production, and the growth of flexible accumulation strategies. In part it reflects the threatened obsolesence of fossil-fuel technologies in the transportation industries. In part it reflects the influx of a less venturesome international capital in metropolitan real estate markets which were once almost exclusive domains of local capital. And in part it reflects the power of neighborhood and community groups to oppose, and thereby impose higher costs on, new development.

In recent decades, attempts were made to purge the pro-growth blocs of labor and progressive elements who had gained new roles during the 1960s. The national urban reforms of the Kennedy–Johnson era were abandoned; labor, African-American and other minority communities were marginalized within the bloc. For example, finance capital willingly entrusted its home town, New York City, to a Democratic Party populist with a solid base in the city's real estate community and white neighborhoods. Edward I. Koch was a mayor for the era of fiscal austerity. He would attempt to transform the historic pro-growth bloc by excising from it the municipal unions, African-Americans and some Latino communities. This would allow for cutbacks in city services that capital wanted in order to keep the lid on labor costs. But while he excised labor and African-Americans, he also exposed the inability of his new coalition members to establish a stable, legitimate government. The racially charged ideological mood created in the atmosphere of fiscal austerity resulted in open racist violence in the white neighborhoods where Koch nurtured his populist base.

The captains of finance were forced to move, cautiously, toward the peaceful retirement of Koch and recomposition of the old pro-growth bloc, this time finding that they had to accept a more organized and militant labor movement and African-American community, and along with it a conception of growth emphasizing equity. They had to deal with a broad coalition that went beyond labor and the minority communities to include peace, environmental, women's and gay movements, and a significant minority of whites. In between mayoral elections, Jesse Jackson had swept New York City voters in the 1988 presidential campaign and helped lay the groundwork for the transformation in city politics. When financier Felix Rohatyn blessed the David Dinkins campaign, the financial club gave its seal of approval to the new reformers. Just to make sure everyone understood who the senior partner in the pro-growth bloc is, and that growth must govern all, only days after the Dinkins inauguration newspapers were filled with stories of a new fiscal crisis, which developers were ready to resolve by offering their new supposedly revenue-generating projects for speedy execution.

The emerging new pro-growth coalitions are taking hold in all major central cities. They seek a wider circle of consensus than the old ones, encompassing segments of labor and minority communities. Labor is being told that construction and service jobs await them in revitalized neighborhoods and the new business centers emerging outside the central cities. African-American groups are being told that new development will not displace minority neighborhoods and there will be no more exclusionary suburban development. Community groups are told they can participate in decisionmaking. The new model is as pro-growth as the old one. The only thing changing is the location of growth, the composition of the pro-growth bloc, and the premises and promises that undergird it.

The new progressive movement

Something new has emerged in the past decade that is causing the pro-growth coalitions to make these promises and modify their plans. It is a new progressive urban reform movement. It is based in minority and working-class neighborhoods, environmental, tenant, peace, women's, gay and lesbian, and other progressive movements. Most of all it is built on neighborhoods and communities and the idea of equality among communities.

The new progressive movement in the United States can trace its legacy to a few experiments in municipal socialism in the early part of the century (see Stave, 1975), the federally-sponsored Community Action and Model Cities experiments of the 1960s, and the movements against displacement of low-income minority neighborhoods in the 1960s and 1970s. Today the progressive movement includes reform governments in a few small municipalities like Berkeley and Santa Monica (California), Burlington (Vermont) and Cleveland (Ohio) (see Clavel, 1986). Among the measures instituted by these governments were tenant protection laws, priority funding of social services for low-income people, and exactions of social benefits from developers. There have also been a whole series of symbolic measures taken by progressive local governments, like the establishment of nuclear-free zones, sister-city relations with developing countries, and resolutions in opposition to U.S. interventionist policies around the globe. Given the extreme financial limitations placed on small local governments, it may well be that their foreign policies have had the most important impact on the body politic.

The strongest expression to date of the new progressive movement has been through the successful candidacies of two African-Americans: Harold Washington in Chicago and David Dinkins in New York. The program of the new progressive reform movement is diverse, but most closely captured in the politics of Jesse Jackson's Rainbow Coalition, which both Washington and Dinkins represent. The program includes an emphasis on racial and social justice, maintenance and restoration of social expenditures cut back during the Reagan austerity years, and protection of minority communities. It is not an anti-growth movement, but one that conditions growth on social and racial justice and, to different degrees, environmental protection. It includes anti-growth philosophies and political forces, but is generally not dominated by them (on New York City, see Arian, *et al.*, 1991; on Chicago, see Bennett, *et al.*, 1987).

The new progressive movement implicitly and explicitly challenges the old reform movements' dictum that local government must operate as a profitable enterprise, a principle that has governed municipal management since the turn of the century. Mayor Harold Washington of Chicago explicitly rejected the notion that the modern metropolis could or should be run according to the principles of private enterprise, or the calculus of accountants. He believed that the bottom line should not be the tax base of a municipality, or its marketability. Instead, he hoped that the metropolis would be seen as a social and cultural value, to be enjoyed and used by all of society, and for which all of society should take political, and fiscal, responsibility. In other words, the use values of the metropolis would be assets to be preserved and developed. This philosophy is more in tune with the actual historical development underway than the antiquated mercantilist notions of the metropolis that predominate in professional management circles.

In Western Europe, the new progressive reform movement emerged in the left-wing socialist and Communist local governments of France and Italy (Szelenyi, 1984), reinforced by the militant housing and community movements of the 1970s. This tradition has favored strategies that give priority to social welfare programs, cooperative economic development, neighborhood preservation and the protection of trade union interests. More recently, it has focused on environmental protection. Witness the growth of the Green parties throughout Europe, and the Red–Green coalitions, with the adoption by many left and labor parties of environmental protection programs. The new trend is especially strong in Italy and Germany, where we find the largest Green parties. In Italy, the large Democratic Left Party (formerly the Communist Party) has a pro-ecology program and is generally allied with the Greens. In Germany, the Social Democratic party, currently a minority, also leans toward a strong pro-environment program and rules in several local governments in alliance with the Greens.

In developing nations, the new progressive movement is emerging in the form of large neighborhood-based social movements of working-class people. In highly urbanized countries like Brazil, Mexico and South Africa, the new urban movements are challenging the dominant approach to metropolitan governance and planning. In many cases in Latin America and Asia, they are winning local elections and key positions within the local state apparatus (see Chapter 7).

All too frequently, local government reform in former colonial nations has been a mechanical copy of reforms in the colonizing nations. For example, the United States is still attempting to market its brand of rational municipal management based on local taxation and loyalty to real estate throughout the developing world. In most places, it encounters central governments committed to strong central controls, and in others local reformers affiliated with national labor movements advocating indigenous solutions to the problems of the metropolis.

Across the globe, new forms of urban democracy are emerging that seek to address metropolitan problems, project a metropolitan vision and incorporate neighborhood planning. Centrally planned economies and some capitalist welfare states offer the metropolitan vision, but fail to incorporate planning at the neighborhood level. On the other hand, while the anti-planning bias and proliferation of exclusionary municipal governments in the United States may mitigate against the pitfalls of grandiose master planning visions, they tend to reinforce inequality. They still rely on an outdated structure of municipal government that does not enfranchise many neighborhoods and communities.

A new form of governance is needed that will make the quality of life now materially possible in the metropolis a reality for all, eliminate structural inequalities, maximize popular participation, and guarantee a pluralism of institutions and parties. It should enfranchise the whole and the parts, the metropolis, neighborhoods and communities, cities and suburbs. It should not be surprising that the forms of governance able to do this are still emerging, since compared to other settlement forms the metropolis is still relatively new and undeveloped, and still accounts for a small proportion of the world's population.

7

COMMUNITY AND METROPOLIS

Planning neighborhood places

Some things go. Pass on. Some things stay. I used to think it was my rememory. You know. Some things you forget. Others you never do. But it's not. Places, places are still there. If a house burns down, it's gone, but the place – the picture of it – stays, and not just in my rememory, but out there, in the world. . . . I mean, even if I don't think it, even if I die, the picture of what I did, or knew, or saw is still out there. Right in the place where it happened.

Toni Morrison, *Beloved*

Although Midaq Alley lives in almost complete isolation from all surrounding activity, it clamors with a distinctive and personal life of its own. Fundamentally and basically, its roots connect with life as a whole and yet, at the same time, it retains a number of the secrets of a world now past.

Naguib Mahfouz, *Midaq Alley*

When the drought comes, and dries up the Uruguay River, the people of Pueblo Federación return to their favorite haunt. Receding, the water leaves a moonscape, and they return.

They now live in a town also called Pueblo Federación, same as their old town was called before the Salto Grande Dam flooded it and left it below water. Not even the cross on the old church steeple can be seen; and the new town is much more comfortable and beautiful. But they return to the old town returned to them by the drought, while they can.

They return and occupy the houses that were their houses and are now the remains of war. They start a fire to make tea and a

Figure 7.1 Corviale.

> roast where grandma died and the first goal and the first kiss occurred, while the dogs dig up the earth in search of bones they hid.
>
> Eduardo Galeano, *El Libro de los Abrazos*

About eight kilometers from the center of Rome, a giant new housing project crowns a hill surrounded by green space. Its name is Corviale, and it has become a favorite target for critics of urban planning since it opened in the mid-1980s. Corviale is a stunning kilometer-long eight-story concrete spine perched on a hilltop. It was supposed to be a model of innovative design in publicly financed housing projects. It was supposed to be an alternative to the tightly packed speculative housing that characterizes Rome's suburbs.

Up close Corviale resembles the worst of public housing in the United States. It has broken elevators, abandoned commercial space and garbage-strewn public spaces. This in a country that can pride itself on some of the best publicly subsidized housing in the West. Only a couple of kilometers away, there is another planned neighborhood, Spinaceto, that is well maintained and functional. Spinaceto has a variety of housing types, pleasant open areas and active services. It is reviled by some, mostly the ideological opponents of publicly financed housing, but it has many admirers as well.

Most Italian planners, and even the project's chief architect, do not dispute the failure of Corviale. The question of interest to planners is *why* it failed. The reasons given by critics include poor design, poor management and poor politics. These all figure in the Corviale disaster, but the main reason is an approach to planning that considers buildings and not communities of people as the central element of urban planning. Corviale became an isolated, self-conscious monument to design principles, an architect's dream, and not a *community* within the metropolis.

Corviale's chief architect, Mario Fiorentino, like many others of his generation, sought to create a new standard for design in a part of Rome that had been dotted with a confusing mix of speculative development (see Acocella, 1981). But the single monumental structure does not lend itself to the social diversity and integration made possible within the many other multi-structure public developments in Rome. Nor does it lend itself to integration with the surrounding urban environment. Spinaceto, on the other hand, has a mix of residential, commercial and open land uses, and is surrounded by similar private development (indeed, the building of Spinaceto increased land values in the surrounding area and facilitated new private housing development). Spinaceto has a mixture of public, cooperative and private housing, and a mix of incomes. In Corviale, there is only housing, and all tenants have low incomes.

The lack of services in Corviale is exacerbated by the low level of car ownership among tenants; they do not have ready access to services elsewhere. The architect deprived the tenants of commercial space by planning an unusual commercial spine on the fifth floor, against all market principles, which summarily failed and closed down. Today the entire fifth floor is padlocked. As with all monumental design, the architect sought to create an environment around his building instead of around the daily lives and historic relations among people.

The common criticisms levelled at Corviale and many other publicly financed developments built as a result of Italy's 1962 urban reform law (Law No. 167) are summed up by one scholar as follows:

> The widespread general criticism centers in fact on the excessive emphasis on the concept of the self-sufficient neighborhood, detached from the built city, and the resulting segregated and formally and functionally separate image of the existing city; on the singular housing type; on the repetition of housing models inspired by the concept of housing as a 'habitat machine'; on

> ignoring the rules of formation and change of the 'existing built environment' (that is, the historic city). . . .
>
> (Ciccone, 1986: 14)

Integrated diversity

Corviale illustrates the problems arising when planners overlook one of the most important principles of progressive planning in the metropolis: *integration* – the comprehensive inter-relationship of social groups, land uses and districts. Traditional architecture and planning seek to achieve integration through design of the physical environment, and typically leave out the people. Integration is not simply a principle of physical design. By focusing on the design project, architects and planners divorce the project from its social environment. They succumb to the fallacy of physical determinism.

A second critical principle of metropolitan planning is *diversity* of social strata and land uses. Unlike simpler settlement forms, the metropolis is characterized by a complex division of territory and the diversity of its parts, which planning may enhance or restrain. Thus, Jane Jacobs considered fundamental 'the need of cities for a most intricate and close-grained diversity of uses that give each other constant mutual support, both economically and socially' (Jacobs, 1961: 14).

Jacobs recognized that integration and diversity were not necessarily contradictory principles. There could be social integration and social diversity at one and the same time. There could be spatial integration even when there is a multitude of land uses, and there could be social integration even when there is a multitude of social groups. Indeed, she advocated the conscious fusion of the principles of integration and diversity in the metropolis to achieve *integrated diversity*.

Jacobs criticized the inability of planners to appreciate diversity among and within neighborhoods. She decried public works projects that ignored metropolitan and neighborhood integration. She had little confidence that planners could achieve integrated diversity.

Herbert Gans, who addressed planners' attempts to achieve socially and spatially diverse communities, shared this skepticism:

> *heterogeneity cannot be achieved until the basic metropolitan area social problem is solved.* This I believe to be the economic and social inequalities that still exist in our society, as expressed

> in the deprivations and substandard living conditions of the lowest socio-economic strata of the metropolitan area population.
>
> (Gans, 1968: 177, emphasis in original)

In other words, simply piecing together separate but unequal parts cannot resolve inequalities. Integrated diversity is not simply a matter of physical design. Planners can arrange the physical proximity of social groups without achieving social integration. After all, slaves usually lived in the same households with their masters.

The U.S. metropolis is an example of diversity without integration. The Soviet metropolis was an example of integration without diversity. The question is to what extent the two societies will change, or be changed, so that planners can realistically achieve integrated diversity.

Neighborhoods as myth and reality

The search for an equitable, integrated approach to planning in the metropolis must be rooted in an understanding of the historic and cultural specificity of neighborhoods. A specific analysis of the livelihood and evolution of every neighborhood is the foundation for an appreciation of metropolitan diversity.

The main component parts of the metropolis are its residential communities – its neighborhoods. The metropolis contains many communities – social groups with common interests such as ethnic minorities, women and real estate developers, for example – but neighborhoods are diverse communities with a fixed spatial location. In many metropolitan areas, neighborhood boundaries coincide with municipal boundaries, but not always. Many suburban neighborhood boundaries coincide with municipal boundaries, but many do not. We should also note that the term neighborhood is often used to refer to central city communities, but suburban communities are also neighborhoods.

There is no universal definition of a neighborhood, only definitions relative to every national and metropolitan context. Neighborhoods are mostly residential quarters defined by a complex of social, economic, physical, institutional and psychological relations and territorial boundaries. There is no fixed criterion of size, or optimum population level, defining a neighborhood; discussions of neighborhoods cover small units of a few thousand population up to

100,000 population and more. Suzanne Keller considers a neighborhood to be any 'physically delimited area having an ecological position in a larger area' (Keller, 1968: 91).

Even where the Central Business District's economic and political hegemony is supreme, or where industry and administrative functions predominate in the local economy, most metropolitan land is used by residential communities. Neighborhoods, especially suburban neighborhoods, increasingly define the spatial organization of the metropolis, especially in the 'post-industrial' era. Social reproduction – how and where people sleep, prepare meals, shop, nurture children and spend their leisure time – is certainly no less important in determining land use in the metropolis than production – how and where people work.

Even while production becomes more and more mobile, neighborhood-based consumption and reproduction are relatively fixed features in the urban landscape. Although manufacturing moves from one metropolitan area to another, neighborhoods in the metropolis stay intact and eventually attract new industries, increasingly in the service sectors. As Susan Fainstein (1987: 323) notes: 'While people retain roots in their communities, economic enterprises no longer do so.'

Even while labor becomes more and more mobile, neighborhoods remain. There is evidence that interactions among neighbors are not necessarily affected by a high level of mobility (see Caplow and Forman, 1974). For example, university towns have a high residential mobility but their residential communities tend to be socially cohesive and stable because the institutional structure is relatively durable.

The cohesiveness and stability of neighborhoods is, of course, relative. Low-income neighborhoods tend to be the most vulnerable to factory closures, service cutbacks and capital flight. The extremely high level of mobility in the U.S. may have some impact on neighborhood stability, but far more important are income and racial inequalities.

The metropolis would not exist without its production units and workplaces, but today its spatial organization and development revolve more and more around the location, dislocation and development of residential neighborhoods, and less and less around production. Production and services are increasingly planned and oriented around residential communities, whereas in the past it was the other way around. Jane Jacobs hit upon this characteristic of the metropolis when she found that today's human settlements that produce to meet internal needs – the great metropolitan areas – tend to grow, and settlements that produce for export decline (Jacobs, 1984).

The permanency of the metropolis is testimony to the objective rise of labor to a more powerful position in the political economy of development. During the rise of industrial capital, investors easily determined the location of production and cities, and attracted labor reserves to them. In the highly metropolitanized nations of the world, investors no longer enjoy such freedom. They may be closing smokestack industries in old cities, but they are constrained by economic calculation to reinvest in metropolitan areas. They are wedded to the metropolis because of its higher level of productivity and efficiency, the diversity of its labor supply, and linkages with other industries and services. In short, they are compelled to invest in what is becoming labor's natural domain, the metropolis.

After the individual and household, the neighborhood is the *basic unit of consumption*, particularly the most socially necessary consumption – the consumption of food and shelter. The quality of neighborhood life – its housing, shopping and recreational facilities – make up in large part the standard of living and level of collective consumption. The quality of life in neighborhoods is a reflection of the social wage and the labor market, which in turn partially determine the location of production units.

The neighborhood is also the basic level for the planning and development of the *social and physical infrastructure* of the metropolis. Clarence Perry was among the first to articulate this principle when he put forth the 'neighborhood unit' concept, a proposal for building communities around elementary schools and a bundle of other basic services (Perry, 1929). The Soviet *rayon* similarly locates the neighborhood as a basic urban unit, but within a much more clearly defined metropolitan structure. The common belief today is that elementary schools, fire and police precincts, clinics and many other facilities function most efficiently when their capacities and service areas are limited to local neighborhoods. The question is whether all neighborhoods can be guaranteed the same basic standard of services, and whether they can respond to the diversity of individual and collective consumption needs. In an unequal social environment, neighborhood units can also be instruments of exclusion and segregation (see Isaacs, 1948).

The neighborhood is one of the most important institutions of *civil society* in the twentieth century. It is the nexus of a whole array of territorially-based civil institutions from block associations to sports leagues, including local booster organizations like the Chamber of Commerce. These institutions arise spontaneously to resolve social needs and problems inadequately addressed by the state and economy, and in reaction to inequalities within the metropolis. The institutions reflect the historical, spatial and cultural differences that define the metropolis. If, as Hegel said, war shows the primacy of the state over civil society, then peace provides an opportunity for the growth of the institutions of civil society.

Because national governments are by definition removed from neighborhoods, metropolitan-level governments are usually weak or non-existent, and municipal jurisdictions often do not coincide with historically developed neighborhoods, civil society has played an increasing role in shaping neighborhoods, if not actually governing them. The institutions of civil society are now a significant social and political reality and potential element in the evolving forms of metropolitan governance and planning. They are not necessarily conservative or progressive, on the political right or left – they can be coopted and manipulated by either – but they are a growing political force.

Neighborhoods are both myth and reality. As *reality*, they are objective phenomena that arise from metropolitan growth within particular economic and historical contexts. Their character and problems reflect the social and economic systems within which they flourish. However, there is also a subjective aspect of neighborhood development. Every neighborhood is, to a greater or lesser degree, a *myth* that evolves in the collective consciousness of its people. Planners may serve that myth or seek to manipulate it, but they cannot avoid it. For example, Harlem is not only an aggregation of African-Americans whose physical environment may be inventoried and described. It is an idea, even an inspiration, that can grip the imagination of poets Langston Hughes and Nicolás Guillén, community activists like Jesse Gray, and revolutionaries like Malcolm X. Sometimes it is so deep it is part of a collective unconscious.

The shape of the neighborhood myth is bound up with the social, economic and physical conditions in which its people live. Thus, it can be an inclusionary myth or an exclusionary myth. It can be a working-class myth or a bourgeois myth. It can be a myth of cultural diversity or a melting-pot myth. It can merge with the myth of the undifferentiated middle class, as in the sprawled North American suburb, but it nevertheless retains a reference, no matter how diffuse, to a local place.

Neighborhoods are not simply transplanted towns or villages. On the surface, they may appear to be reproductions of traditional pre-metropolitan communities, but they are not, either objectively or subjectively, urban villages, as suggested by Herbert Gans' popular term (1962). Nor are they the same as the alienating residential quarters of nineteenth-century industrial cities like London and Birmingham. They are neither the *Gemeinschaft* nor *Gesellschaft*, village and town, of nineteenth-century urban sociology. They do not have the organic geographical, social or economic independence of the village or town. Herein lies the main contradiction of metropolitan neighborhoods. They exist within a large, diverse and interdependent metropolis, yet their very survival depends on the maintenance of distinctions with the rest of the metropolis – a consciousness of their territory, their own place.

Kevin Lynch (1960) was among the first to analyze the strong connection between the physical structure of neighborhoods and subjective images. Logan and Molotch (1987) show how the sense of place that people have at the neighborhood level is founded on real relations among people and social groups, and is also a distinct objective and political element.

These conceptions of place diverge sharply from the traditional geometric project approach to design handed down by generations of architects and planners who have yet to come to grips with the complexities of the new metropolitan life. The traditional approach is at best three dimensional, yet urban places cannot be properly understood without the fourth dimension, the dimension of individual and collective consciousness.

Our sense of urban territory is based on local neighborhood experience passed down from one generation to the next and intertwined with the cultural history of nations and peoples. Morris and Hess (1965: 1) state that 'a sense of neighborhood haunts our history and our fondest memories.' Neighborhood boundaries are usually not defined in a precise way either geographically or politically, but they exist in the collective consciousness of people. Objective and subjective aspects are interrelated such that the idea of one's neighborhood is an objective factor in the survival and development of the neighborhood. The myth of my neighborhood can help to rally my neighborhood and thus to create it or save it.

Collective neighborhood consciousness is not simply based on shared experiences in a common territorial space. It may also be based on social homogeneity and economic (especially property) interests.

In reviewing studies of suburban communities, Gans (1968: 153) concluded that relations of friendship are more likely to be founded on social homogeneity than propinquity. To the extent that neighborhoods are socially homogeneous, they may engender greater social interaction, but there is no inherent magic in neighborhoods that creates cohesiveness or social integration.

Matthew Crenson's (1983) study of Baltimore suggests that neighborhood consciousness is based on social diversity, not homogeneity. Crenson found that diversity and inequality produced neighborhood political activity and that 'cohesiveness' within neighborhoods was mostly a myth. Crenson believes that diversity and not homogeneity is the basis for achieving equality among neighborhoods, and that by seeking a false cohesiveness the result will be greater social exclusion and segregation.

Neighborhood, households and women

In the metropolis the potential exists as never before for the gradual transfer of the functions performed by households to the institutions of the neighborhood and the local state. This process is more than just a potential; it is already under way. In many places it is being pressed by women and women's organizations, but even where it is not the process is occurring. The neighborhood will never replace the individual or household as the most fundamental unit of natural and social reproduction, at least for the foreseeable future. But it can and does play an increasingly important role in reproduction.

Smith and Tardanico (1987: 100) consider the household to be the missing link between work, community and collective action. For them,

> the household is the basic microstructure of social, economic and political life in which working life and residential community life are experienced as united in family life. The activities of people living in households interact with the activities of people living in other households. Some of these interactive processes of communication lead to the formation of informal social networks within classes. Ultimately, household activities are the basic elements of group and class formation in any social system. . . . Furthermore there is a connection between the process of capital accumulation and the social functions performed in households. Households physically reproduce

> labor power by generating income through a combination of formal and informal sector employment and state transfers and services; they also are the primary unit engaged in socialization and cultural reproduction, including socialization to the societal contradictions which give rise to social movements.

The overall trend in the metropolis, especially in the most developed metropolitan areas, is toward the socialization of labor-force reproduction. Socialization means that the activities that once took place in the private household now take place outside the household within larger social institutions, many of them neighborhood-based institutions. This process is comparable to the socialization of production that occurred when manufacturing moved from the individual craft shop to the factory. It is spurred by the accumulation of capital and the commodification of consumption activities, and is also associated with the phenomenon of conspicuous consumption (see Chapter 2). Socialized reproduction includes both publicly owned and privately owned enterprises; it does not necessarily mean direct government involvement. Since the neighborhood is the primary locus of labor-force reproduction, the socialization of reproduction creates the possibility for conscious neighborhood planning.

The socialization of reproduction is evident in the growing role of neighborhood-based services once performed in the household: food preparation (fast food), clothes repair, education, child care and leisure-time activities. In U.S. suburbs, for example, the food court at the shopping mall provides meals for many families and is also a favorite social gathering spot for young people. Childcare services expanded dramatically when the proportion of women in the labor force grew. And voluntary clubs and associations offer activities for people of all ages.

Socialized reproduction is not universal, and reflects general social inequalities. Countervailing tendencies are many. In Third World countries where there is a large 'informal sector', households continue to perform many services, and even take on manufacturing activities that were once previously socialized. In the United States, there is an ideological throwback to the nuclear family by the far right, which would have women perform their traditional roles in the household even when they are part of the labor force. There are technological innovations, such as VCRs and home computers, that make the prolongation of the isolated household possible. However, in the U.S. the nuclear family no longer predominates; only one-fourth of all households are traditional nuclear families with husband, wife and children. In sum, the traditional equation of family and household no longer holds.

In many parts of the world, there has always been a close relationship between household and neighborhood, a characteristic inherited from the town and village. For example, in southern Europe and Latin America, the *piazza* and the street are not simply transit ways between home and work but places where children play, older people gather to socialize and lovers rendezvous. Public space is organically related to, and an extension of, the family's private space. This explains in part the receptivity to large investments in public infrastructure, especially mass transit. This organic relation between public and private space, however, has been broken in much new metropolitan development, which follows the atomizing process of the North American auto-centered model.

It is no accident that women have been among the most active pioneers seeking to understand the laws governing the relationship between public and residential space – women like Jane Jacobs (1961), Dolores Hayden (1984) and Roberta Gratz (1989). Their sharp critiques of the (largely male-dominated) planning profession place a particular value on the centrality of street life and public spaces in densely developed cities like New York. We might also note that another woman, Suzanne Keller (1968), was the first to compile a comprehensive book on the neighborhood, and Rosabeth Moss Kanter (1972) wrote a classical work about intentional communities.

Jane Jacobs approaches planning as first and foremost a neighborhood question, not a matter of physical design or rationalization of metropolitan growth. She notes how a densely developed and integrated urban environment with a variety of architectural types and social groups, like her neighborhood of residence, Greenwich Village, makes a vibrant community. Her neighborhood nurtures diversity and human interaction. It is designed (or allowed to grow with less conscious intervention) on the human scale, to protect people and public spaces through such institutions as the 'eyes on the street.' This contrasts with the approach of planning to protect buildings and increase the value of property, the sterile formalism and dull uniformity of technocratic models, and the grand visions of the master builder.

Following in the tradition of Jane Jacobs, Roberta Gratz (1989) highlighted the failure of planners to give priority to preservation of housing and neighborhoods. Dolores Hayden (1984) shows how the

North American model of sprawled urban growth creates neighborhoods based on fear and alienation, especially for women and children. The model of the single-family dream house on an individual plot of land promotes the segregation of workplace and residence, alienates people from public spaces, and isolates the household – women in particular. Hayden seeks a new integration of workplace and residence, workplace issues and housing issues, workplace activities and household activities. She would reconstruct domestic spaces so that they may incorporate socialized reproduction – childcare and recreation, for example. She would also domesticate urban space, so it is not just a way station between workplace and home but an integral part of the neighborhood. She seeks improved public transit and community facilities, free from intimidating symbols of male domination such as billboards with half-clad women. Whereas the suburban American Dream seeks to separate production and reproduction, her vision is of a neighborhood that integrates them.

Women's appreciation for the value of neighborhoods reflects their collective experience with the problems of childrearing, housework, shopping, local transportation, crime, leisure time and schools. From this understanding has come a challenge to the leaders of both capital and labor – capital where it has erected private spaces and destroyed public spaces for profit, and labor where it has stood by passively or hailed the process. Women play no small role in the growing movement within labor away from narrow issues of wages and working conditions toward issues of work time, public services and the quality of life.

If, as Andre Gorz (1985) suggests, the key to the progressive project today is separating work time from pay, then it may be possible in the future for more work to be household-based and neighborhood-based. People will have a guaranteed income and thus be able to undertake community service and creative enterprises integrated with their livelihood at home. However, this kind of future seems to be workable only in the most developed economies with high levels of social surplus. In the majority of metropolitan areas around the world, such integration of workplace and residence is not a hallmark of development but a product of household survival strategies in the face of poverty, unemployment and the restructuring of capital (Smith and Tardanico, 1987).

The considerable role played by women in housing and community movements, especially since the 1960s, follows from the objective tie between reproduction in the family and neighborhood welfare. However, it has grown most rapidly when women move rapidly into the organized labor force (see Nash and Safa, 1976). As women across the globe have joined the ranks of wage labor, neither the state nor the marketplace have adequately assumed the functions previously performed by women in the household. Instead, women have often been forced into the double shift. They perform socialized labor at the workplace and privatized labor at home. They are therefore in the forefront of the movements for the socialization of reproduction, the strengthening of civil institutions and the expansion of the local state. In short, women are key actors in movements over collective consumption (see Datta, 1990: 56).

The management and planning of the metropolis across the globe tends to be in the hands of male-dominated professions. The eventual political change that brings planning for neighborhoods into central focus may also prompt the concurrent end of the male monopoly over urban management, metropolitan planning and decisionmaking.

Neighborhood movements and community control

Neighborhood institutions vary as widely as the national cultures and regions that shape them. They include single-issue organizations, spontaneous protest groups, civic societies, social clubs, tenant and homeowner associations, block associations, and many more. Since the upsurge of urban protest movements in the 1960s and 1970s, there has been much discussion about the role, existing and potential, of urban movements in bringing about a more just and equitable society. Marxists have asked whether neighborhood movements tend to be progressive and pro-working class (see Castells, 1977; Cowley, 1977; Katznelson, 1981). Unfortunately, the issue has often been posed in the form of a crude dualism that equates the working class with struggles at the point of production and *déclassé* community groups with struggles over consumption. In the classical economic determinist view, working-class movements are only those tied to unions, wages and working conditions, and community-based movements are seen as inherently populist in character. Reality in the post-industrial metropolis is much more complex than this rigid paradigm suggests.

Neighborhood movements, like all other movements, have no inherent *a priori* class character or political direction. They are not necessarily, or mostly, progressive or democratic, as suggested by

Figure 7.2 Community and tenant movements have played a key role in neighborhood preservation, development and planning.

Harry Boyte, who hails neighborhood activism as 'a democratic awakening' (see Boyte, 1979). They may be faithful allies of profit-motivated real estate developers or advocates of neighborhood preservation. They may be allied with right-wing political parties as well as the left. They may champion racist exclusion as well as fair housing, the privatization of space as well as its socialization (for views that contrast with Boyte, see Jones, 1979; also Kling and Posner, 1990).

Similarly, organizations of labor are by no means uniformly radical, militant, left or progressive. For example, the building trades in the United States have systematically excluded African-Americans from higher paying jobs, supported conservative candidates, and have done little to support fair housing in the exclusive suburbs where they live, or programs for the improvement of low-income communities.

Spontaneous and militant protest movements, especially at the local level, are often of short duration and their creative energy does not necessarily become institutionalized (see Piven and Cloward, 1971; 1979). The most recent protests over consumption, spontaneous rebellions begun in the late 1980s against economic austerity measures instituted at the behest of the International Monetary Fund, have not evolved into powerful organized forces able to install alternative governing regimes. As John Walton (1987) suggests, the most significant impact of these protests has been to change the terms of international debt negotiations. The terms of debt repayment may ultimately have an effect on the quality of life at the neighborhood level. However, the pressures of individual daily survival do not enhance the power and stability of poor people's movements in low-income neighborhoods – one reason for the growth of broader political parties that draw on the support of popular movements.

The largest and most powerful neighborhood movements have formed national coalitions, allied themselves with political parties, and projected a broad vision of social change encompassing the metropolis and beyond. It is perhaps no coincidence that many of these movements have arisen in Latin America, the world region with both large metropolitan areas and mass urban poverty. The neighborhood movements of Mexico, Brazil, Chile, Peru and Venezuela have been particularly important social and political forces (Castells, 1983: 173-212; Fox, 1989). Let us briefly examine the case of Mexico.

Millions of Mexicans belong to neighborhood organizations that seek improved housing and services and at the same time project a revolutionary transformation of society. While not characterized by any single ideology, they have generally advocated progressive, democratic and socialist visions of social transformation. They have coalesced in the Coordinadora Nacional del Movimiento Urban Popular (CONAMUP), the Coordinadora Unica de Damnificados (CUD), and the Asamblea de Barrios, led by the popular Super Barrio, and many other local coalitions. Many of these forces backed Cuauhtémoc Cárdenas' left-leaning Frente Nacional Democrático in the 1988 elections. The Frente was a coalition of left forces independent of the ruling Institutional Revolution Party (PRI), and lost the election by a narrow margin amidst Frente charges of electoral fraud by the governing party. Many neighborhood activists were prominent in the electoral campaign, and some were candidates in local elections.

The neighborhood movements arose spontaneously in the 1960s and 1970s in response to government attempts to evict shantytown dwellers, particularly the poor families that had invaded vacant suburban land to establish homes for themselves. The new neighborhoods organized to fight for basic services like water and sewer lines, and to get the government to recognize them as legitimate participants in urban politics. The Mexico City neighborhood and tenant movements played a key role in organizing to press for government action to rebuild areas devastated by the 1985 earthquake.

The Mexican neighborhood movements have been the main conduit for the entrance of large numbers of women into Mexico's civic and political life. For example, CONAMUP sponsored three National Women's Conferences, and all of the major coalitions have spawned independent women's initiatives for women's rights.

The PRI and the National Federation of Popular Organizations (CNOP) with which it is allied have attempted with limited success to coopt the neighborhood movements. The PRI-backed Mexican Labor Federation (CMT) has maintained its traditional focus on workplace organizing and a non-confrontational relation with the government. The PRI government, committed to IMF-imposed austerity measures, has not been able to meet neighborhood demands. Thus, the neighborhood movements have had to coalesce at the metropolitan and national level, and join with other unrecognized social movements and opposition political parties, or face dissipating local protests with no prospects for concrete results (Velázquez, 1989).

In the U.S., the majority of neighborhood organizations are suburban and loyal to the suburban ethic of exclusion. However, neighborhood movements that seek to defend against displacement have generally tended to promulgate a progressive neighborhood myth based on inclusion and the equalization of living conditions within the metropolis. They include the movements against the destruction of neighborhoods by the federal urban renewal and highway programs, gentrification, plant closures and institutional expansion. They include the tenant movement, and major protests such as the Harlem rent strikes of the 1950s, San Francisco's International Hotel struggle of the 1970s, and the Cedar-Riverside (Minneapolis) struggle of the 1970s. (For others see Hartman, Keating and LeGates, 1982.)

Movements against displacement are defensive in character. But struggles against displacement invariably steer toward the more affirmative objective of *community control* – achieving direct control over territory and urban space. For communities that have been historically dispossessed, empowerment means control over the future of the urban environment in the same way that more privileged groups have control over their neighborhoods. It means preserving neighborhood institutions and the built environment from the ravages of the real estate market, drugs, unemployment, poverty and disease. Contrary to the theory of 'internal colonialism' which sees African-American neighborhoods as colonies (see Turner, 1970), it is not a matter of a nation controlling its borders, but a people and culture organized as a neighborhood controlling its urban places. (See Plotkin, 1987; for a compendium of anti-displacement strategies, see National Urban Coalition, 1978; 1979; Hartman, Keating and LeGates, 1982.)

Community control does not necessarily conflict with social or racial integration. The fight for a decentralized school system in New York City during the 1970s was a key part of the fight for community control by African-American and Latino neighborhoods. Many advocates of racial integration opposed community control of the schools, saying it would reinforce racially separate neighborhoods. The opposition failed to see the link between the defense of territory (community control) and the fight against racial exclusion. School decentralization promised to give African-American and Latino neighborhoods the same kind of control over their neighborhood schools that whites had enjoyed *de facto* for decades. With a local power base, African-Americans would be in a better position to fight for equal treatment in education and all other areas, and by achieving equality with respect to white neighborhoods the whole logic of exclusion would fall apart. The philosophical arguments against school decentralization also obscured another underlying power struggle: the top-heavy central Board of Education bureaucracy and the American Federation of Teachers, both of which were disproportionately white, saw community control as a threat to their jobs and power (see Altshuler, 1970).

NEIGHBORHOOD STRATEGIES

In the following sections, we will examine various strategies for the preservation, development and planning of neighborhoods: decentralized government, community economic development, housing development, homeownership and planned neighborhood development.

Decentralized government

One of the strategies for enhancing neighborhood power is decentralized governance within the metropolis. Because of its size and complexity, governing the metropolis through a single centralized administration obviously produces inefficiencies if not inequalities.

Yet there are surprisingly few examples of decentralized metro governments. Some of the more interesting experiments have been in Europe (Sharpe, 1979). In the United States, most of the decentralization experiments have been within large central cities. However, these are exceptions and not the rule.

The problem is not simply one of formally decentralized management and planning but above all political power and economic equality. For example, New York City has one of the most developed forms of decentralized governance in the world, but it is more form than substance. Its neighborhoods are vastly unequal in economic and political power, so decentralized governance reinforces these inequalities. There are five boroughs and fifty-nine community districts in New York City. Each of the community districts has a community board with up to fifty voting members. However, the community boards play a very limited role within government. Their members are appointed, serve on a voluntary basis, and their decisions are non-binding. Each board has no more than a few paid staff members and handles an area containing roughly 100,000 people. The boards face an often faceless municipal bureaucracy of over 300,000 employees, powerful independent authorities and elected officials. (This decentralized form only works within the five boroughs of New York City, which account for less than half of the metropolitan area's total population.)

There has been much talk about creating neighborhood government fully entrusted with municipal powers (Kotler, 1969), but this remains a dream. In most metropolitan areas, political power continues to reside in larger municipalities, powerful government agencies, political clubs and elite civic or business groups formally outside local government. Where it is more dispersed, exclusionary and wealthier neighborhoods often have disproportionate powers (see Chapter 6).

There is no consensus on the appropriate size for decentralized metropolitan government. There can probably be no universal standard and the optimum size varies with the regional and national contexts. The number arrived at in New York City and Moscow is 100,000 – the average size of a New York City community district, and the neighborhood planning unit (the *rayon*) in the former Soviet Union.

There is also a need for a smaller unit of political organization – perhaps on the block level. The Chinese, Cuban and Nicaraguan revolutions established active citizen committees at the block and district levels that perform a variety of functions including provision of social services, discussion of national legislation, public security and cultural activities. Unofficial civic organizations perform these functions in other countries. Official promotion and financing of organizations at this level, while running the risk of bureaucratization, offer greater opportunities for grass roots participation and the equalization of resources.

Community development and neighborhood preservation

Local actions to preserve neighborhoods and enhance neighborhood control often end up promoting growth and the growth ethic. Indeed, this may be their inevitable direction. Once an immediate threat of displacement is over, and a measure of political independence has been secured, the next question is how to improve the neighborhood. To paraphrase Martin Luther King, Jr., once the lunch counters in the South were desegregated in the 1950s, the next question was how to afford a meal. However, unlike the opening up of lunch counters and the assertion of community control, the basic economic solutions to neighborhood problems cannot be found at the local level.

Community movements thus give rise to institutions and strategies dedicated to *community development*. Community development may be equated with either *growth* or *preservation*, but usually it is associated with the former. Community development that combines growth and preservation is more likely to encourage integrated diversity. Community development that is supported by metropolitan-wide and national policies is more likely to encourage integrated diversity.

One of the most popular local growth strategies is *greenlining* – the promotion of individual home ownership and local real estate capital, in direct opposition to redlining by banks, insurance companies and government. This method seeks to cure the patient by injecting him or her with the same bacteria that caused the illness in the first place. It may benefit individual homeowners and provide some stability for a portion of the community, but it cannot stop neighborhood displacement, because it does not resolve the problems of inequality in wages and living standards, and cannot deal with the mobility of capital and labor in a *laissez-faire* system (see Angotti, 1979; Edel, Sclar and Luria, 1984).

Another local strategy is to promote employment, particularly through local development corporations (LDCs). Neighborhoods press for more jobs, or try to encourage a more cooperative and equitable system of production within their borders. The problem with the job promotion approach is that economic growth must occur within the context of the national and international division of labor. Local neighborhoods can only fight among each other for a finite number of jobs, and cannot eliminate the structural reserve army of labor, the source of unemployment in the first place. Indeed, capital depends on this fight because it allows individual investors to negotiate lower infrastructure costs and higher subsidies with local governments. As Logan and Molotch (1987: 89) note, 'local growth does not make jobs: it only distributes them.'

Housing development

One of the most common strategies for neighborhood preservation is the development of low-income housing. Government-financed housing programs have traditionally been a major instrument for community development and preservation. Public housing tends to be relatively permanent and stable, and in neighborhoods with many low-income households public financing is the only practical option for housing significant numbers of people. In the U.S., less than 3 per cent of the total housing stock is subsidized for low-income people. European and socialist countries, where housing has been considered a basic human right, have maintained relatively high levels of government subsidy to low-cost housing. In Great Britain 35 per cent of the total housing stock is publicly owned, 33 per cent in Denmark and Germany, 15 per cent in France and over 80 per cent in the former Soviet Union (Fuerst, 1974: 178; see also McGuire, 1981). However, large public housing sectors have also supported development of relatively free-market capitalist countries, such as Hong Kong and Singapore (Castells, Goh and Kwok, 1990). In most of the countries with significant public housing, subsidized units tend to be available to households from diverse income strata and there is less of a stigma attached to living in public housing.

Neighborhood-based strategies to provide local alternatives to low-income housing have come more and more into focus during the Reagan/Thatcher years of social austerity, when public subsidies were cut drastically. Housing advocates were faced with an old dilemma more sharply than ever before: to fight for more federal assistance or to build local alternatives. While the two are not mutually exclusive, the question of which ought to be the priority inevitably arises.

Figure 7.3 Eldridge-Gonaway apartments in Oakland, California, forty units of low-income housing built by Oakland Community Housing, Inc., financed with tax increment funds from Oakland's City Center development. Use of the City Center fund for low-income housing was a concession won by community groups after years of opposition to downtown displacement and urban renewal.

The problem is that local responses to the housing crisis, while often important and innovative, have not been able to come close to matching even the modest levels of federal government assistance, nor are they likely to in the future. Furthermore, the most successful ones have relied heavily on what little government assistance there is. This does not mean local initiatives should not be undertaken, but that they should be seen as part of a wider national effort to restore government involvement in housing. A national perspective is especially important since the conservative cutbacks are usually justified by glorifying local alternatives.

One local alternative is to turn to municipal government to finance low-cost housing. This has proven successful in countries where municipalities have substantial authority for raising their own revenues, as in the United States. But given the continuing problem of homelessness in the U.S., local initiative has obviously been insufficient to solve the problem. The main local effort has been through alternative financing of housing – through tax increment funds,

Figure 7.4 Renewal without removal. Bologna (Italy) used government subsidies to renovate historic housing in the central city. This enabled tenants with low and moderate incomes to remain in their neighborhoods while the old housing was preserved.

revenue bonds, revolving loan funds, housing trusts, etc. Most of these funds are used to write down loans for moderate-income owner-occupied housing, often for young first-time homebuyers.

Only in a few central cities most affected by the homelessness crisis have local government policies been used to develop low-income rental housing. For example, San Francisco's zoning ordinance requires downtown developers to contribute to a housing trust fund or provide low-income housing in order to have their office and luxury projects approved (between 1981 and 1985, commitments were made to build 3,800 units, about 70 per cent of them for low- and moderate-income housholds). New York City has perhaps the most ambitious local production program, which at its peak produced several thousand units a year in new and rehabilitated units.

But production alone cannot solve the problem, especially if the new low-income units are built in exchange for new office and luxury construction, because the new downtown construction only reproduces the original problem and fuels the process of gentrification and displacement. Production of more low-income units in central cities also allows suburbs that continue to exclude low-income housing to continue to evade any responsibility for the housing crisis.

Non-profit housing providers, or community development corporations (CDCs), run up against the same set of problems as local government. Many of them came out of the housing protest and community movements of the 1960s and 1970s, and have come to fruition in the 1980s, following a long and arduous path to legitimacy, and long after the spontaneous movements behind them ebbed into inactivity. They face all the problems that local government housing providers do, but have a further liability. They usually have no independent revenue-generating capacity, and have to rely on voluntary contributions, local government subsidies or public–private partnerships in which they usually wind up the junior partners (see Squires, 1989). Local non-profits are widely promoted in the United States as an alternative to public housing. (See Urban Planning Aid, 1973, for a critique of CDCs.)

According to a 1988 national survey, over 600 local development corporations have produced a grand total of approximately 125,000 housing units, 65 per cent of which were the result of rehabilitation rather than new construction (cited in Keating, Rasey and Krumholz, 1990). This may be contrasted with the over four million units of low- to moderate-income housing that were assisted by government subsidies over the last forty years. Most of the non-profit units benefited from some form of government assistance and, despite the intentions of autonomy, are threatened with extinction by federal austerity measures. This has led as many as 10 per cent of them to undertake for-profit ventures (Meyers, 1985).

Successful non-profits face a serious but inescapable dilemma when increasing land values in the neighborhood make the exchange value of the land far superior to its use value for low-income housing. This makes the prospect of resale or refinancing very attractive. Probably the most dramatic case is the 15,000-unit Coop City in New York City which, along with hundreds of other developments financed through the state and city Mitchell–Lama moderate-income housing program, faces the threat of gentrification. After twenty years, cooperators were given the option of selling their moderate-income units and cashing in on the sizeable equity gains built up over the years due to rising land values. Many cooperators wanted to go to Florida with their pockets full of windfall profits, while Coop City would be gentrified and a major source of moderately priced rental housing would disappear from the New York scene. Under pressure from tenants and community groups, resale restrictions in most Mitchell–Lama projects were extended.

In sum, local housing policy needs to focus not only on the growth in the number of housing units, but on the preservation of neighborhoods.

Homeownership

Homeownership is often promoted as a strategy for both neighborhood development and preservation. After the Reagan-era disengagement of government from public housing subsidies, many have been promoting 'affordable' low-rise owner-occupied housing as a local alternative to high-rise public housing (see Schwartz, Ferlauto and Hoffman, 1988). To be sure, homeownership is usually associated with stability of tenure and quality of maintenance. However, there is no reason why tenants cannot also be guaranteed stability of tenure, as they are in many cities with strict rent laws and relatively dormant real estate markets. As tenants in luxury units will testify, there is no reason why apartments cannot be well maintained. Public housing tenure is stable; but for legislative prejudice against providing decent services to poor people, and inept authority administration, it could be well maintained too. If low-income tenants themselves were given greater responsibility for maintenance, the results might be quite different.

This is not to say that homeownership is necessarily either a desirable or undesirable form of housing tenure, only that it is not a panacea. In response to proposals by Proudhon and others advancing homeownership as a solution to the housing question, Friedrich Engels insisted that changing the title to the property would make little difference as long as the economic condition of workers remained the same:

> The whole conception that the worker should *buy* his dwelling rests again on the reactionary basic outlook . . . according to which the conditions created by modern large-scale industry are morbid excrescences. . . . [This is] nothing but an idealized restoration of small-scale enterprise, which has gone and is still going to rack and ruin.
>
> (Engels, 1975: 31)

While some Marxists have interpreted this to be a wholesale rejection of homeownership, many others continue to advocate homeownership under specified conditions. For example, in Cuba practically all housing is owned by its occupants, while urban land is publicly owned.

There is also a tendency to identify mortgage debt as the source of the housing problem. This reflects a misplaced concern for the problems of homeownership. In his critique of capitalist homeownership, Michael Stone (1986) proposes the elimination of mortgage debt and the eventual socialization of housing. Stone implies that mortgage debt is the source of the housing crisis, and the recent Savings and Loan failures would seem to vindicate this view. However, it would be well to take note of Engels' critique of earlier proposals to reduce interest rates and get the banks out of housing:

> The interest on loaned *money* is only a part of profit; profit, whether on industrial or commercial capital, is only a part of the surplus value taken by the capitalist class from the working class in the form of unpaid labor. . . . As far as the distribution of this surplus value among the individual capitalists is concerned, it is clear that for industrialists and merchants who have in their businesses large amounts of capital advanced by other capitalists the rate of profit must rise – all other things being equal – to the same extent as the rate of interest falls. The reduction and final abolition of interest would . . . do no more than re-arrange the

> distribution among the individual capitalists of unpaid surplus value taken from the working class. It would not give an advantage to the worker as against the industrial capitalist, but to the industrial capitalist as against the rentier.
>
> (Engels, 1975: 34)

Mortgage lending may be socially useful if it allows people with limited means to pay for their housing over a long period of time. The broader regime of private accumulation and consumption, not mortgage debt, is the source of the housing problem.

Homeowners may be working class in relation to the main means of production but they are petty capitalists in relation to urban property. Homeowners are preoccupied with mortgage debt because it makes their position relative to capital as a whole unfavorable. No matter that they are mortgaged up to their teeth; they still own the property. And when they fight with uncharacteristic militancy to protect their property values, they can put themselves on the other side of progressive forces. It is not the same as the fight of homeowners displaced by gentrification, redlining, and municipal service cutbacks. Some homeowners thrive on the misery of others; other homeowners know nothing but misery.

In sum, it is not the cost of housing by itself, to either tenants or homeowners, that is the problem. It is not a tenant question or a homeowner question. If rents and mortgage payments went down, so would salaries. If working people had to pay nothing at all for housing, it would produce:

> a depression of the value of labor power and will therefore finally result in a corresponding drop in wages. Wages would thus fall on an average as much as the average sum saved on rent, that is, the worker would pay rent for his own house, but not, as formerly, in money to the house-owner, but in unpaid labor to the factory owner for whom he works. In this way the savings of the worker invested in his little house would in a certain sense become capital, however not capital for him but for the capitalist employing him.
>
> (Engels, 1975: 48–9)

In countries where there are both tenants and homeowners, tenants are usually more vulnerable to loss of place, and a disproportionate number of low-income people are tenants. In the United States, over half of the minority population are tenants compared to only 40 per cent in the general population. Almost two-thirds of minorities in central cities are tenants. For this reason, tenant movements in central cities are usually focused on broader issues of social equality and neighborhood preservation.

Many low-income tenants have more in common with low- and moderate-income homeowners than with upper-income tenants. Therefore, a strictly tenant-based movement often focuses narrowly on organizing individual buildings instead of the neighborhood. Narrowly based tenant organizations are often pitted against homeowners. Significantly, the strongest rent control ordinance in the U.S., in Berkeley, California, is the product of a relatively strong coalition between tenants and low-income homeowners.

The most stable foundation of the tenant movement has been low-income tenants (for a useful tenant history, see Lawson, 1986). In this light, proposals that hinge on organizing 'a new generation of middle-class tenants' (see Atlas and Drier, 1986: 378) have not come to fruition. Still, low-income tenants can make little progress without joining together with other tenants and low-income homeowners.

Tenant organization and tenant movements began like the trade union movements, around demands against property owners. There is fundamentally no difference in that both tenant and trade unions demand changes that seek to improve the standard of living of their members, one through wage and benefit increases and the other through rent control and improvements in housing conditions. Trade unions tend to be more consistently working class in composition, but both tenant and trade union organizations can serve conservative labor aristocrats, technocrats and corporativistic-minded workers.

Rent controls can contribute to the stabilization of neighborhoods, but cannot solve the housing problem. Rent controls exist at the national level in several European countries, where they have stabilized tenancy and, in combination with ongoing subsidies for low-income housing, slowed down gentrification. However, rent control within an unequal land and labor market produces new forms of inequality. It pushes housing investors towards building housing for sale, and many existing owners toward the use of informal and illegal methods of eviction, and toward conversion to condominiums. It does not necessarily affect land values, but may encourage landlords to take their biggest profits out of long-term equity gains (when they sell the property) rather than short-term operating income.

Local rent control is even more precarious than national rent control. In the United States over 250 municipalities have some form of limitation on annual rent increases ranging from the very mild to the strict. This includes central cities such as New York City, San Francisco, Washington, and a gradually expanding number of inner-ring suburbs in the tri-state area around New York City. Rent control was given a big boost by the February 26, 1986 ruling of the U.S. Supreme Court in *Fisher v. City of Berkeley*, which upheld the constitutionality of the country's strongest rent control law. Local rent controls, which usually include eviction controls, have had some effect on slowing rent increases and preventing displacement, but when only some jurisdictions in the metropolis have rent control they have allowed landlords in other jurisdictions to charge higher rents. (For a review of rental legislation, see Gilderbloom and Appelbaum, 1988.)

Some housing advocates have proposed the 'de-commodification' of housing (see Achtenberg and Marcuse, 1986) as a strategic objective. The decommodification of housing would entail some form of public or social ownership of housing and democratic, neighborhood-based control. But even if the entire *housing* stock were to be socially owned (through public or non-profit institutions), there could still be a private *real estate* market and its attendant inequalities, and a private *labor* market with its inequalities. Going back to Engels' argument with Proudhon, *it matters little who holds title to the property*, as long as the fundamental underlying economic relations are exploitative.

The ultimate goal ought to be *decommodification of labor* and elimination of the reserve army of labor (though not necessarily the elimination of competition among labor). This would create the basis for the provision of stable and affordable housing for all.

Another key goal should be the *social ownership of land*. This would facilitate the planning of integrated diversity. Under these conditions, a regulated market could exist in housing, but differences in price would not reflect underlying economic and spatial inequalities.

In conclusion, the improvement of housing at the neighborhood level is severely constrained by economic considerations, particularly the land and labor markets, and national and metropolitan policies. Strictly local neighborhood action is not sufficient for neighborhood improvement. An isolated solution to the housing problem is not possible.

Neighborhood development strategies must focus on both growth and preservation to achieve integrated diversity. However, integrated diversity based on economic equality cannot be attained by individual neighborhoods acting in isolation from one another, or without equitable metropolitan-wide and national social policies.

Perhaps realization of the enormous dimension of the underlying problems leads many activists at the neighborhood level to give up, or to turn inward. As Manuel Castells (1983: 331) states, 'when people find themselves unable to control the world, they simply shrink the world to the size of their community.' The slogan of the Greens, 'Think globally, act locally,' is an acknowledgment of the immensity of the underlying problems and a call to action. However, local action can also obscure the necessity of metropolitan-wide and national reforms.

Planned neighborhood development

Neighborhood planning is limited when a strictly localist outlook prevails. Its prospects are even dimmer if there is no comprehensive metropolitan or national planning. Nowhere has planning at the three levels been perfectly balanced and coordinated. However, there are many experiments and models around the world that suggest that it is possible to integrate neighborhood and metropolitan planning and make progress toward integrated diversity. These experiments include the development of entirely new neighborhoods – often called 'new towns in town' or 'cities within cities.' They also include comprehensive preservation of existing neighborhoods.

The theory and practice of neighborhood-level planning is still in its infancy and has only recently begun to generate attempts at a comprehensive analysis (see Werth and Bryant, 1979; Bayor, 1982; Clay and Hollister, 1983; Banerjee and Baer, 1984; Hester, 1984; Rohe and Gates, 1985). Socially conscious planners seek to understand how neighborhoods with which they work function and develop. Metropolitan-level planners especially need this insight since they are not as closely tied to individual neighborhoods. Such an understanding cannot be simply deduced from the general laws of metropolitan development, but needs to be induced at the street level, through direct interaction with people and places in neighborhoods. Planners at the metropolitan level can chart out the basic policies of infrastructure and development strategy in the region. However, unless these policies are informed by a knowledge and understanding of neighborhoods, they will end up destroying neighborhoods, as the sad history of 'urban renewal' testifies. When they use only the metropolitan-wide vantage point to plan, they either run up against intensive community opposition or, even worse, are able to implement community-destructive plans.

In sum, only planning at the metropolitan level can establish policies of equitable distribution of services and infrastructure throughout the region. Only planning at the neighborhood level can establish inclusionary plans for neighborhood preservation and development. The secret to metropolitan planning is the successful integration of the two.

Lauchlin Currie proposed a multi-level approach to neighborhood development. He outlined a program for 'cities-within-cities' with populations ranging from 400,000–500,000. They would be planned to integrate many smaller neighborhood clusters, each of which would be only large enough for pedestrian circulation to be adopted as the main transportation mode. Single ownership of land, local administration, local employment, coordination with regional authorities and within a national urban policy are other key elements of his proposal (Currie, 1976).

In a similar approach, urban designer Eduardo Lozano (1990) recommends:

> The establishment of metropolitan authorities is essential for dealing with the regional issues subsumed under growth and development policies: the location of main centers of employment, the extension of public transit lines, the banking of land for integrated housing. These authorities should have the power to override local municipalities in order to zone land in strategic centers and corridors (though not in all the metropolitan areas), to plan main public transportation lines, to acquire land for housing, and to apportion property tax revenues from and to the various municipalities. . . . The reconstruction of urban communities must stress diversity, as both a social and visual characteristic, and ensure a variety and equality of social groups, so that no one can exert undue dominance. It must stress a rich urban mixture with economic and political balance, an environment that is resilient, enjoyable, and fair.

A necessary but not sufficient condition for this kind of integrated approach to neighborhood planning is *social ownership of land.* As long as land is owned and controlled by forces that have no vested interest in integrated metropolitan and neighborhood development, particularly real estate speculators and absentee landlords, neighborhoods will have little power to determine the use of their land. Social ownership of land – by locally-based cooperatives and land trusts, public authorities, or in some cases permanent owner-occupiers restricted by controls on sales – can better guarantee neighborhood stability. The variety of social forms of ownership is limitless, but the common element is removal of land from the real estate market.

Outright government ownership of all land does not by itself insure integration or diversity; government may even thwart implementation of these principles. The Soviet model of consumption was based on collective principles. However, it was flawed because it sought to apply the same approach to consumption as it did to production: serial reproduction on a large scale of standardized units. This yielded factory-issue neighborhoods, mechanical reproductions of standard urban designs. The result was an alienation of individual and household from urban places. The wide boulevards and massive infrastructure, though generally favoring public over private means of transportation and use, have no spirit of local history or place.

There are many alternatives to such a monolithic system of state ownership, such as Henry George's idea of taxing away all profits on private land (George, 1880), long-term leaseholds, municipal ownership, non-profit land trusts and limited-equity cooperatives. Capital, and capitalism, can do very well without a real estate market because land does not produce new capital; the land market only redistributes capital. At the same time, socialism can do very well without state ownership of land, because the state is not the only alternative to capital. Therefore, there are no fundamental economic reasons why land cannot be socially owned in a variety of forms under capitalist, socialist or mixed economic systems.

Integrated neighborhood planning can occur within the context of certain types of private ownership. Single private or mixed public–private ownership of large tracts of land provide better conditions for socially conscious planning than the anarchy of many competing owners. Columbia, Maryland and Reston, Virginia are two examples of comprehensively planned, relatively integrated private neighborhoods. However, they are still isolated models in the suburban sea of exclusion, mostly upscale in social composition and careful to limit their low-income populations. Their greatest accomplishments are in physical design, with a variety of housing types and innovative systems of pedestrian ways, though they are still locked into the disintegrating use of the automobile as the main means of transportation (see Tennenbaum, 1990).

More comprehensive experiments may be found in the suburbs of some Scandinavian and other European metropolitan areas. Satellite suburbs of Stockholm, Helsinki, Oslo, Paris, Rome and Berlin offer prominent examples of integrated diversity. In most cases, there is social ownership of land, significant public planning at the metropolitan level, integration of new neighborhoods with other parts of the metropolitan area, and the use of public policy to achieve some level of social and spatial diversity. These new neighborhoods generally go beyond the small-scale Garden City villages that also dot the European landscape, and the first generation of British New Towns. They are decidedly urban, not suburban, in form and function.

The planning of Stockholm's satellite neighborhoods was made possible by municipal ownership of almost all land outside the city center. Public ownership was instigated not, as might be suspected, by the Social Democratic government after World War II, but by the Swedish Crown in 1904, which permitted the local government to acquire on favorable terms agricultural land surrounding the city. Social Democratic legislation in 1953 gave the municipality powers of expropriation, and by the mid-1960s the city government owned 70 per cent of all suburban land (Odmann and Dahlberg, 1970; Stockholm City Building Department, 1972).

A comprehensive master plan for Stockholm was adopted in 1952. The plan projected the development of suburban satellite neighborhoods linked with the central city by mass transit. The first generation of these neighborhoods, including Farsta and Vallingby, had populations of some 50,000. More recent ones are larger and include a wider variety of services. They have retained a relatively dense and compact urban form encouraging mass transit, pedestrian and bicycle use instead of auto use. This policy uses suburban land more efficiently and allows for preservation of the plentiful steep slopes and natural waterways that characterize the Stockholm metropolitan landscape.

Stockholm's planned neighborhoods were facilitated by public ownership of most developable land by a single entity, a relatively homogeneous population, an unusually high standard of living with low unemployment and a minimal labor reserve, national social legislation providing significant income and housing subsidies, including rent subsidies to low-income people, and a tradition of cooperation among labor and capital and among local governments. While these conditions may be unique, they are also present to varying degrees in many other countries. The contrast with the competitive capitalist regime of the U.S. is, however, dramatic.

In Italy, the problematic models like Corviale, which we examined at the beginning of this chapter, may be contrasted with many positive examples of integrated diversity. Rome's new neighborhoods were made possible by national legislation in 1963 that allowed municipalities to acquire large suburban tracts of land at below-market prices. The land was developed in accordance with the 1962 comprehensive plan for the Rome Comune, which includes within its boundaries practically the entire metropolitan area. Some forty-three different projects ranging in size from 1,000 to 30,000 population were undertaken, though only a few of them, like Spinaceto, are large enough to have a distinct neighborhood character (Comune di Roma, 1981; 1986).

Spinaceto now has a population of about 25,000. Approximately 40 per cent of the population live in public housing, 40 per cent in private housing and the rest in cooperatives. Shopping and other services are located within the neighborhood, along a central spine that gives the neighborhood a distinct spatial definition (Crescenzi, 1965). There is a variety of relatively dense and compact housing types, but no sharp differences between the quality of the housing complexes. National housing policies and rent control provide stability of tenure as well as social diversity. New industrial and commercial enterprises have opened near Spinaceto, which is conveniently located next to Rome's ring road. Surrounding private land has been developed with housing which is mostly compatible in scale and character with Spinaceto.

Some of the problems of Spinaceto's development signal more pervasive difficulties that Rome planners, and all new neighborhood planners, have had in achieving integrated diversity. The first problem was the extensive delay in completing construction, which began in 1966, and especially the lag in providing services (see Angotti, 1977b: 38–42). Originally designed to connect with a new suburban rail line, Spinaceto is still accessible only by bus and auto, which are the main means of transportation. The limited size of Spinaceto does not permit the development of a full range of services within the neighborhood, although this is less of a problem because it is readily accessible to, and integrated with, surrounding neighborhoods.

We could examine hundreds of examples of integrated diversity in metropolitan areas like Stockholm and Rome. They are not Utopias or

Figure 7.5 Crystal City, Virginia, a suburban center outside Washington, D.C., and an example of planning for private spaces.

ideal neighborhoods. They are made possible by the relatively abundant economic surplus of developed nations. It is not only their design that makes them liveable, but *mainly* the economic and social welfare of their inhabitants, which is a reflection of economic wealth and is guaranteed by liberal national welfare policies. They all face the planning problems of service lag, linking workplace and residence, and the design of public spaces that reinforce historic urban traditions. But they all follow the principles of integration and diversity.

One of the most important lessons of new neighborhood development has been that planners cannot bring together workplace and residence through physical design. While neighborhoods can be planned with their own industry and services, it is not necessarily useful, either from an individual or social standpoint, that people should work in their own neighborhoods, as Currie and others have suggested. The metropolis offers many opportunities for mobility and interchange between districts; it is often easier for people to change their jobs than their place of residence, and there is no reason why employment mobility should be curtailed. The separation of workplace and residence, both physical and psychological, is an inevitable consequence of the modern division of labor, labor mobility and

Figure 7.6 Stockholm. The Husby neighborhood, planned with a density similar to that of the central city. Non-auto circulation predominates in Stockholm's suburbs.

transportation mobility. Neighborhood planning that begins and ends with the objective of physically reuniting workplace and residence does not take into account the complexity of the metropolis. It can also lead to the isolation and segregation of neighborhoods.

Recent experiences in which workplaces in homes have been made possible by new electronic technology, particularly micro-computers, faxes and fiber-optic communications, are limited. These opportunities are still only available to a small upper-income stratum of the labor force. In many cases, the physical union of workplace and residence has been historically associated with low-wage labor and poverty – piecework at home in the garment industry, for example. Informal sector employment throughout the Third World is filled with examples of workplace/residence integration, which are symptomatic of household survival strategies and labor paid below the average wage level.

Integrated diversity can best be achieved by conscious comprehensive planning at both metropolitan and neighborhood levels. If physical form is the guiding element, however, in new neighborhood development, as with Corviale, integrated diversity is bound to suffer. On the other hand, if profit is the guiding element, new

Figure 7.7 Warsaw. An historic plaza reconstructed after World War II from plans hidden away during the Nazi occupation of Poland.

neighborhoods will be like miniatures of Las Vegas or Disneyland theme parks in which everything is a commodity up for sale, including the most basic human activities.

John Short, in *The Humane City* (1989), criticizes planning for cities as if profits, professionals and some people were all that mattered, and calls for 'cities as if people matter.' He would go beyond the existing liberal, conservative, capitalist and socialist models, and explore new mechanisms for planning through community ownership, cooperatives of labor and the evolving mixed world economy.

The Disneyland approach to neighborhood development is gradually replacing the traditional approach of sprawled suburban subdivisions in the U.S. Its objective is to promote the sale and consumption of commodities by creating compact comprehensively planned, inwardly oriented communities. Disneyland planning creates neighborhoods in the image of the Central Business District.

A prototype of the new Disneyland neighborhood is Crystal City, Virginia, only a ten-minute drive from downtown Washington, D.C. It is a dense business/residential neighborhood with none of the advantages of high density and central location. Served by freeways and an extensive street system, it has no street life. It is served by a

Figure 7.8 The Bazaar in the medieval center of Cairo. The open marketplace is one of the oldest public places. Planning should seek to preserve and enhance it.

mass transit line, but mass transit in the D.C. area is of the new generation (of which systems in San Francisco and Atlanta are also a part) that principally serve a small proportion of suburban commuters rather than the masses. All activity in Crystal City is turned inward, within the private buildings and private spaces. It is an eery secretive environment appropriate to the nearby Pentagon and CIA, whose functionaries are among Crystal City's executive customers. Crystal City is an attempt to create an urban environment outside the central city, but it only reproduces the prevailing suburban isolation. Crystal City has no public places.

All conscious plans should be judged by the extent to which they facilitate the real social interaction of people in metropolitan space. The neighborhood is the locus of the most universal form of interaction – the sharing of *public places*. As shown by William H. Whyte (1988), planning for urban places is too often a second thought in the process of constructing buildings and public works. Planners design spaces without consideration for the practices, needs and desires of the urban public. Whyte shows how small urban spaces, often unplanned, can be intensively utilized, and monumental planned spaces tend to be abandoned.

Figure 7.9 Piazza Navona (Rome), a popular public place only recently reclaimed from the invasion of the central city by motor vehicles. Credit: Bruce Dale.

Whyte, a social commentator and not an architect or planner, takes the kind of approach to planning that is sorely lacking in many of even the best neighborhood plans, including Farsta and Spinaceto. He first of all observes people in busy urban spaces, especially streets, and then asks how to plan. He looks first at how neighborhoods work, and then asks how to make them work better. This reminds me of an experienced campus planner who once told me how he planned campus walks by seeding open spaces and then paving the areas that were worn down by traffic. Whyte is engaged in the daunting task of retrofitting urban design monsters ranging from imposing downtown skyscrapers with threatening ground-floor facades to suburban shopping centers surrounded by deserted parking lots.

Whyte, Jane Jacobs and many other critics of official planning understand that the density and scale of building is not necessarily the problem in planning public places. Some of the most exciting and intensively utilized public places – such as Piazza Navona in Rome, Washington Square in New York, and the Grand Place in Brussels – are surrounded by densely developed neighborhoods.

Busy, crowded streets, not expansive commercial strips, are the most sought after public places (see Rudofsky, 1969). It is possible to have vibrant and secure public places in densely developed urban environments as well as in low-density environments, in both cities and suburbs. It is possible to break down the distinctions between central city and suburbs and create urban places throughout the metropolis. The central city will never lose the value of centrality, but suburbs do not have to forfeit urbanity.

In sum, planned neighborhood development can produce new communities with many of the amenities of central areas and without some of their problems. Neighborhood preservation strategies can help eliminate some of the problems of existing neighborhoods. However, development and preservation can also reproduce inequalities unless they are part of broader metropolitan-level planning. If they follow the principles of integrated diversity, they can help ameliorate inequalities. The comprehensive planning of new neighborhoods can take place without threatening the preservation of old neighborhoods.

8

CONCLUSION

The future of the metropolis

While there are some important exceptions, metropolitanization has been a companion of economic development in the twentieth century. This is a basic law of the metropolis and the basis for a progressive, humane approach to city planning. When the metropolis – all of it, central city and suburb, and all metropolitan areas, rich and poor – is understood as a necessary and beneficial accompaniment to social progress and not an endemically flawed artifact, city planning can begin to contribute to that progress. The emergence and dominance of the metropolis in the twentieth century is indisputable, and the import of this social fact has not yet been fully comprehended by urban analysts and planners.

In the broadest sense, the metropolis is the means by which the historic differences between city and countryside are being eliminated. As Hans Blumenfeld noted, it is neither city nor countryside, but a synthesis of the two. The inequalities between city and countryside are being replaced by inequalities between metropolitan areas, and within them.

It would be precipitous, however, to declare the complete supremacy of the metropolis and the end of the historic division between city and countryside. By the end of the twentieth century, the century of the emergence of the metropolis, the majority of the world's population will still reside outside the metropolis, and almost half will still live in towns and rural areas. Metropolitan planning, particularly in the poorest countries of Africa and Asia, can achieve very little until the newly independent nations overcome economic dependency and underdevelopment, and establish a framework of national planning within which metropolitan areas and their neighborhoods may plan. It is hard to imagine how the potential of urban planning can be fully realized, however, without a new international economic order.

North America is the most metropolitanized region of the world. However, the majority of the world's population lives in the developing nations of Africa, Asia and Latin America, where most metropolitan areas are located. The dependent metropolis of the South, not the developed metropolis of the North, is the most prevalent, and will become even more prevalent in the future. The dependent metropolis is based on inequality, and its proliferation means the reproduction of inequality.

The dominance on a world scale of the U.S. model of metropolitan planning threatens the global environment and can only heighten social inequalities. In the U.S., there needs to be a fundamental change in the regime of accumulation that gives greater priority to long-range economic planning, economic and racial equality, and economic and spatial integration. There needs to be a new way of thinking about the metropolis that is urban and not anti-urban. The U.S. has yet to develop a conscious national urban policy that acknowledges the metropolis as a whole, and gives sufficient aid to central cities, mass transit, and urban services. Federal policy should promote neighborhood organizations and neighborhood preservation, and the ownership and management of housing by stable neighborhood-based institutions and individuals. Land-use regulations should be inclusionary and not exclusionary, and planning should move beyond zoning. Reform of the U.S. planning model along these lines would have a powerful impact around the world, especially among those countries dependent on the U.S.

The demise of the Soviet Union has led to widespread repudiation of the Soviet approach to metropolitan planning. However, it would be well to strive for a balanced assessment of Soviet city planning that takes into account both contributions and shortcomings. Planners in the former Soviet republics who are able to maintain their balance in this time of revolutionary ferment are advocating the introduction of regulated markets, greater diversity in urban design, decentralized planning, the empowerment of neighborhoods and communities, and an end to Utopian formalism. They see the need for a variety of forms of land and housing ownership. Many continue to uphold the principles of integrated planning and social equality. It remains to be seen whether these planners will have their way.

In centuries to come, even the metropolis may wither away as technological and economic conditions change. Perhaps the 'megalopolis' of Jean Gottmann (1961), the 'conurbation' of Patrick Geddes (1915), or the 'informational city' of Manuel Castells (1989) will become the standard. Perhaps the polycentric form of more limited metropolitan development characteristic of large portions of Europe and the Soviet Union will predominate over the sprawled metropolis of North America.

The restructuring of production, new communications technology and the revolution in information systems all make possible a drastically reordered metropolis, and ultimately a new international settlement order, perhaps highly decentralized and dispersed. However, these phenomena are still very limited. With over half the world living in poverty, a fourth living in housing unfit for human habitation, and millions still dying from starvation, such a future is very far off.

NOTES

1 THE CENTURY OF THE METROPOLIS

1 In the tables, I have generally used Habitat estimates for 1990 (Habitat, 1987), which appear to correct many of the problems with the European data found in other United Nations reports. The latest United Nations data (United Nations, 1989) were also consulted, and in the few cases where the actual population data exceeded the Habitat estimates (frequently in Asian cities and metros), the higher figure was used. The Habitat estimates are far from perfect, and the astute reader may find them far off the mark in some cases, but in the aggregate they appear to be within the actual order of magnitude on a global scale.

3 THE DEPENDENT METROPOLIS: DEVELOPMENT AND INEQUALITY

1 Especially in its earliest stages, dependent capitalism relies on primitive accumulation. Primitive accumulation involves 'the expropriation of the immediate producers, i.e., the dissolution of private property based on the labour of its owner' (Marx and Engels, 1968: 235). Small-scale peasant and artisan production is shattered, encouraging migration to the metropolis.

While aspects of pre-capitalist modes of production, and the early capitalist mode of production dependent on primitive accumulation, continue to exist today in mature capitalist systems, their development is qualitatively shaped by the dominant mode of production. This is contrary to the structuralist analysis that focuses on concurrent 'modes of production.'

4 THE SOVIET METROPOLIS

1 *Subbotniks* were volunteer work parties organized to work on Sundays on public works projects. The Stakhanovite movement was a movement among factory workers promoting emulation of highly productive workers. It was named after the worker who spurred the movement.

APPENDIX

Metropolitan areas (over one million population) in 1990

AFRICA

Country	*Metropolis*	*Population ('000)*
ALGERIA	Alger	3,380
ANGOLA	Luanda	1,900
BENIN	Cotonou	1,512
CAMEROON	Douala	1,030
EGYPT	Alexandria	3,350
	Cairo-Giza-Shubr	8,640
ETHIOPIA	Addis Ababa	1,880
GHANA	Accra	1,630
GUINEA	Conakry	1,470
IVORY COAST	Abidjan	1,189
KENYA	Nairobi	1,700
MOROCCO	Casablanca	3,240
	Rabat-Sale	1,150
MOZAMBIQUE	Lourenço Marques (Maputo)	1,470
NIGERIA	Ado-Ekiti	1,955
	Ibadan	1,296
	Ila	1,096
	Ilorin	2,542
	Kaduna	1,084
	Lagos	4,790
	Mushin	1,447
	Ogbomosho	1,028
LIBYA	Tripoli	1,720
SENEGAL	Dakar	1,480
SOUTH AFRICA	Cape Town	2,250
	Durban	1,444
	East Rand	1,563
	Johannesburg	2,310
	Port Elizabeth	1,116
	Pretoria	1,010

AFRICA – *Continued*

Country	*Metropolis*	*Population ('000)*
SUDAN	Khartoum	1,800
TANZANIA	Dar es Salaam	1,820
TUNISIA	Tunis	1,630
ZAIRE	Kanaga	2,113
	Kinshasa	3,300
ZAMBIA	Lusaka	1,030
TOTAL AFRICA		73,365

ASIA

Country	*Metropolis*	*Population ('000)*
AFGHANISTAN	Kabul	1,297
BANGLADESH	Chittagong	2,322
	Dacca	6,530
	Khulna	1,497
BURMA	Rangoon	3,170
CHINA	Anshan	1,694
	Beihai	3,362
	Beijing	9,290
	Benxi	1,412
	Botou	1,593
	Chang Chun	5,705
	Changsha	2,460
	Changzhou	1,787
	Chengdu	4,025
	Chongqing	6,511
	Dalian	4,619
	Dangdong	2,574
	Fushun	2,045
	Fuxin	1,693
	Fuzhou	1,651
	Guangzhou	5,670
	Guiyang	1,350
	Hangzhou	5,234
	Harbin	2,790
	Hefei	1,706
	Hohhoit	1,146
	Huaibei	1,308
	Huainan	1,519
	Huangshi	1,069
	Jilin	3,974

ASIA *-Continued*

Country	*Metropolis*	*Population ('000)*
	Jinan	3,376
	Jinzhou	4,448
	Kunming	1,976
	Lanzhou	3,899
	Liaoyang	1,612
	Liupanshui	2,107
	Luoyang	1,900
	Luta	2,177
	Nanchang	2,471
	Nanjing	3,682
	Nanning	1,504
	Paotao	1,654
	Pinxiang	1,189
	Qingdo	4,205
	Qiqihar	1,209
	Shanghai	11,960
	Shaoxing	1,091
	Shenyang	5,055
	Shijiazhuang	1,332
	Suzhou	1,029
	Tai'an	1,275
	Tainjin	7,790
	Taipei	2,910
	Taiyan	2,205
	Tangshan	1,408
	Tsinan	1,730
	Tsingtao	1,836
	Urumqi	1,173
	Wenzhou	5,948
	Wuhan	4,273
	Xi'an	2,912
	Yichun	1,167
	Yingkow	1,612
	Zaozhuang	2,704
	Zhengzhou	1,943
	Zibo	2,198
	Zigong	1,673
HONG KONG	Victoria	5,620
INDIA	Agra	1,087
	Ahmedabad	3,760
	Bangalore	5,140
	Bhopal	1,016
	Bombay	11,790
	Calcutta	12,540
	Coimbatore	1,896

ASIA *-Continued*

Country	*Metropolis*	*Population ('000)*
	Delhi	9,130
	Dhanbad	1,525
	Hyderabad	3,700
	Indore	1,137
	Jabalpur	1,127
	Jaipur	1,485
	Kanpur	2,350
	Lucknow	1,379
	Madras	6,030
	Madurai	1,791
	Nagpur	1,753
	Poona	2,580
	Salem	1,048
	Surat	1,281
	Tiruchirapalli	1,349
	Ulhasnagar	2,866
	Vadodara	1,096
	Varanasi	1,029
	Visakhapatnam	1,154
INDONESIA	Bandung	2,227
	Jakarta	9,480
	Medan	3,010
	Palembang	1,080
	Semarang	1,269
	Subarabaja	3,823
IRAN	Esfahan	1,232
	Karaj	1,796
	Mashad	1,260
	Shiraz	1,217
	Tabriz	1,046
	Teheran	8,331
IRAQ	Baghdad	8,203
	Basra	1,440
ISRAEL	Tel Aviv-Yafo	1,616
JAPAN	Chiba	1,231
	Fukuoka	1,176
	Hiroshima	1,055
	Kawasaki	1,106
	Kitakyushu	2,096
	Kyoto	1,528
	Nagoya	2,330
	Osaka-Kobe	10,686
	Sapporo	1,568
	Tokio-Yokohama	23,372
KOREA (DPRK)	Hamhung	1,156

ASIA -*Continued*

Country	*Metropolis*	*Population ('000)*
	Pyongyang	1,755
REPUBLIC KOREA	Inchon	1,387
	Kwangchu	1,003
	Masan	1,365
	Pusan	4,900
	Seoul	11,660
	Taegu	2,163
LEBANON	Beirut	2,060
MALAYSIA	Kuala Lumpur	1,750
PAKISTAN	Faisalabad	1,104
	Hyderabad	1,137
	Karachi	8,160
	Lahore	4,350
	Lyallpur	1,987
	Multan	1,030
	Rawalpindi	1,396
PHILIPPINES	Dagupan	1,089
	Davao	1,152
	Manila	8,260
	Quezon City	1,420
SAUDI ARABIA	Jeddah	1,772
	Riyadh	2,380
SINGAPORE	Singapore	2,700
SYRIA	Aleppo	1,405
	Damascus	1,640
THAILAND	Bangkok-Thonburi	7,380
TURKEY	Ankara	3,630
	Istanbul	5,476
	Izmir	1,658
VIETNAM	Danang	3,864
	Haiphong	1,279
	Hanoi	2,571
	Ho Chi Minh City	3,420
TOTAL ASIA		472,932

LATIN AMERICA

Country	*Metropolis*	*Population ('000)*
ARGENTINA	Buenos Aires	11,710
	Cordoba	1,285
	Rosario	1,122

LATIN AMERICA *-Continued*

Country	*Metropolis*	*Population ('000)*
BOLIVIA	La Paz	1,210
BRAZIL	Belem	1,357
	Belo Horizonte	3,890
	Brasilia	2,400
	Cabo	1,577
	Curitiba	3,772
	Fortaleza	2,422
	Goiania	1,788
	Novotguacu	1,325
	Porto Alegre	3,180
	Recife	3,040
	Rio de Janeiro	11,370
	Salvador	2,650
	Santos	1,139
	São Paulo	18,770
CHILE	Santiago	4,550
COLOMBIA	Baranquilla	1,775
	Bogotá	5,270
	Cali	2,402
	Medellin	3,601
CUBA	La Habana	2,040
DOMINICAN REPUBLIC	Santo Domingo	2,170
ECUADOR	Guayaquil	1,638
	Quito	1,220
GUATEMALA	Guatemala	1,460
MEXICO	Ciudad Juarez	1,006
	Guadalajara	3,200
	Leon	1,077
	Mexico	20,250
	Monterrey	3,010
	Puebla	1,260
PARAGUAY	Asuncion	1,350
PERU	Lima/Callao	6,780
PUERTO RICO	San Juan	1,816
URUGUAY	Montevideo	1,248
VENEZUELA	Caracas	4,180
	Maracaibo	1,295
	Valencia	1,135
TOTAL LATIN AMERICA		147,740

NORTH AMERICA

Country	*Metropolis*	*Population ('000)*
CANADA	Montreal	2,907
	Toronto	3,330
	Vancouver	1,572
UNITED STATES	Atlanta	2,561
	Baltimore	2,280
	Boston	4,059
	Buffalo	1,294
	Charlotte	1,065
	Chicago	8,111
	Cincinnati	1,690
	Cleveland	2,766
	Columbus, Ohio	1,299
	Dallas	3,655
	Dayton	1,222
	Denver	1,847
	Detroit	4,611
	Fayetteville	1,053
	Fort Lauderdale	1,772
	Fort Worth	1,185
	Hartford	1,043
	Houston	3,635
	Indianapolis	1,335
	Jacksonville	1,000
	Kansas City	1,621
	Las Vegas	1,076
	Los Angeles	13,075
	Louisville	1,120
	Memphis	1,009
	Miami	2,912
	Milwaukee	1,611
	Minneapolis	2,615
	New Orleans	1,334
	New York	17,968
	Norfolk	1,310
	Oklahoma City	1,020
	Oxnard/Ventura	3,279
	Philadelphia	5,833
	Phoenix	1,885
	Pittsburgh	2,316
	Portland, Oregon	1,364
	Providence	1,186
	Sacramento	1,291
	St Louis	2,536
	St Petersburg	1,034

NORTH AMERICA *–Continued*

Country	*Metropolis*	*Population ('000)*
	Salt Lake City	1,041
	San Antonio	1,275
	San Bernardino	1,240
	San Diego	2,269
	San Francisco	5,878
	San Jose	2,477
	Seattle	2,343
	Tampa	1,914
	Washington, D.C.	3,563
TOTAL NORTH AMERICA		144,687

EUROPE

Country	*Metropolis*	*Population ('000)*
AUSTRIA	Wien	2,044
BULGARIA	Sofia	1,600
CZECHOSLOVAKIA	Praha	1,191
DENMARK	Kobenhavn	1,352
FRANCE	Lille	1,298
	Lyon	1,388
	Marseille	1,304
	Paris	8,680
GERMANY	Berlin	3,290
	Frankfurt	1,965
	Hamburg	2,190
	Munchen	2,180
	Rhein/Ruhr	9,252
	Stuttgart	1,916
GREECE	Athinai	3,027
HUNGARY	Budapest	2,085
IRELAND	Dublin	1,100
ITALY	Florence	1,169
	Genova	1,230
	Milano	7,530
	Napoli	4,150
	Roma	3,750
	Torino	2,400
NETHERLANDS	Rotterdam	1,028
POLAND	Gdansk	1,073
	Katowice	3,460
	Lodz	1,086

EUROPE *-Continued*

Country	*Metropolis*	*Population ('000)*
	Warszawa	1,662
PORTUGAL	Lisboa	2,200
	Porto	1,315
ROMANIA	Bucuresti	2,370
SPAIN	Barcelona	3,240
	Bilbao	1,821
	Madrid	4,950
	Sevilla	1,394
	Valencia	1,766
SWEDEN	Malmö	1,103
	Stockholm	1,540
SWITZERLAND	Zürich	1,004
UNITED KINGDOM	Birmingham	2,890
	Glasgow	1,846
	Leeds/Bradford	2,000
	Liverpool	1,557
	London	10,400
	Manchester	2,510
	Newcastle	1,110
YUGOSLAVIA	Beograd	1,212
TOTAL EUROPE		120,628

USSR

Country	*Metropolis*	*Population ('000)*
	Alma Ata	1,357
	Baku	1,777
	Chelyabinsk	1,194
	Dnepropetrovsk	1,226
	Donetsk	1,171
	Erevan	1,148
	Frunze	1,055
	Gorky	1,554
	Kazan	1,228
	Kharkov	1,691
	Kiev	2,920
	Krasnoyarsk	1,176
	Kuibyshev	1,478
	Leningrad	5,430
	Minsk	1,975
	Moskva	9,540

USSR *-Continued*

Country	*Metropolis*	*Population ('000)*
	Novosibirsk	1,659
	Odessa	1,318
	Omsk	1,287
	Perm	1,253
	Rostov-na-Donu	1,138
	Saratov	1,080
	Sverdlovsk	2,360
	Tashkent	1,210
	Tbilisi	1,210
	Ufa	1,222
	Volgograd	1,255
	Voronezh	1,101
	Yerevan	1,368
	Zaporozhye	1,115
TOTAL USSR		54,496

OCEANIA

Country	*Metropolis*	*Population ('000)*
AUSTRALIA	Adelaide	1,133
	Brisbane	1,211
	Melbourne	3,230
	Perth	1,296
	Sydney	3,930
NEW ZEALAND	Auckland	1,015
TOTAL OCEANIA		11,815

Source: Habitat, 1987; United Nations, 1989 (adjusted estimates)

BIBLIOGRAPHY

Aaron, Henry J. (1972) *Shelter and Subsidies: Who Benefits from Federal Housing Policies*, Washington, D.C.: The Brookings Institution.

Achtenberg, Emily Paradise and Peter Marcuse (1986) 'Toward the decommodification of housing', in Bratt, Hartman and Meyerson: 474–83.

Acocella, Alfonso (1981) *Complessi Residenziali Nell' Italia degli Anni '70*, Firenze: Alinea.

Adams, Gordon (1984) 'The Department of Defense and the military-industrial establishment: the politics of the iron triangle,' in Frank Fischer and Carmen Sirianni, Eds., *Critical Studies in Organization & Bureaucracy*, Philadelphia: Temple University Press.

Alexander, Alan (1982) *Local Government in Britain Since Reorganization*, Boston: Allen & Unwin.

Altshuler, Alan (1966) *The City Planning Process: A Political Analysis*, Ithaca, NY: Cornell University Press.

—— (1970) *Community Control: The Black Demand for Participation in Large American Cities*, Indianapolis: Bobbs-Merrill.

—— *et al.* (1984) *The Future of the Automobile: The Report of MIT's International Automobile Program*, Cambridge: MIT Press.

American Society of Planning Officials (1960) *Cluster Subdivisions*. Information Report No. 135, Chicago.

Amin, Samir (1976) *Unequal Development: An Essay on the Social Formations of Peripheral Capitalism*, New York: Monthly Review.

Andrusz, G.D. (1987) 'The built environment in Soviet theory and practice,' *International Journal of Urban & Regional Research*, 11, 4, December: 478–99.

Angotti, Thomas (1973) 'Planning for regional waste water systems,' *Growth and Change*, 6, 2, April: 36–42.

—— (1977) 'The housing question: Engels and after,' *Monthly Review*, 29, 5, October: 39–51.

—— (1977b) *Housing in Italy: Urban Development and Political Change*, New York: Praeger.

—— (1978) 'Planning and the class struggle? Radical planning in the post-Banfield era,' in Harvey A. Goldstein and Sara A. Rosenberry, Eds., *The Structural Crisis of the 1970s and Beyond: The Need for a New Planning*

Theory, VPI & SU: 209–15.
—— (1979) 'A critical assessment of current approaches to housing finance,' *The Black Scholar*, 11, 2, November/December: 2–12.
—— (1981) 'The strategic questions for the housing movement: racism and displacement,' Paper presented to conference on New Perspectives on Urban Political Economy, American University. May: 42pp.
—— (1983) 'New directions in Cuban housing,' *New World Review*, 51, 1, January/February: 12–16.
—— (1986) 'The housing crisis of the 1980's: the alternatives to austerity and the limitations of localism,' unedited version (26pp.) of 'Housing strategies: the limits of local actions,' *Journal of Housing*, 43, 5, September/October: 197–206.
—— (1987) 'Urbanization in Latin America: toward a theoretical synthesis,' *Latin American Perspectives*, 53, 14, No. 2, Spring: 134–56.
—— (1990) 'The housing question: progressive agenda and socialist program,' *Science and Society*, 54, 1, Spring: 86–97.
Arian, Asher, Arthur S. Goldberg, John H. Mollenkopf and Edward T. Rogowsky (1991) *Changing New York City Politics*, New York: Routledge.
Armstrong, Warwick and T.G. McGee (1985) *Theatres of Accumulation: Studies in Asian and Latin American Urbanization*, London: Methuen.
Ashton, Patrick (1978) 'The political economy of suburban development,' in Tabb and Sawers, Eds.: 64–89.
Atlas, John and Peter Drier (1986) 'The tenants' movement and American politics,' in Bratt, Hartman and Meyerson: 378–97.
Auletta, Ken (1982) *The Underclass*, New York: Vintage.
Babcock, Richard F. (1969) *The Zoning Game*, Madison: University of Wisconsin Press.
Bach, Joachim (1974) 'The German Democratic Republic,' in Arnold Whittick, Ed., *Encyclopedia of Urban Planning*, New York: McGraw-Hill, 446–58.
Baldassare, Mark (1986) *Trouble in Paradise: The Suburban Transformation in America*, New York: Columbia University Press.
Bandyopadhyay, Pradeep (1982) 'Marxist urban analysis and the economic theory of rent,' *Science and Society*, 46, 2: 162–96.
Banerjee, Tridib and William C. Baer (1984) *Beyond the Neighborhood Unit: Residential Environments and Public Policy*, New York: Plenum Press.
Banfield, Edward (1958) *The Moral Basis of a Backward Society*, New York: Free Press.
—— (1970) *The Unheavenly City*, Boston: Little, Brown.
Barringer, Felicity (1991) 'Census reflects restless nation,' *New York Times*, December 16: A16.
Bater, James H. (1980) *The Soviet City: Ideal and Reality*, London: Edward Arnold.
Bayor, Ronald H., Ed. (1982) *Neighborhoods in Urban America*, New York: Kennikat Press.
Bell, Gwen, Ed. (1976) *Strategies for Human Settlements: Habitat and Environment*, Honolulu: University of Hawaii Press.
Bellush, Jewel and Murray Hausknecht, Eds. (1967) *Urban Renewal: People, Politics and Planning*, Garden City, NY: Anchor Books.

Benevolo, Leonardo (1967) *The Origins of Modern Town Planning*, Cambridge, Mass.: MIT Press.

Bennett, Larry, Gregory D. Squires, Kathleen McCourt and Philip Nyden (1987) 'Challenging Chicago's growth machine: a preliminary report on the Washington administration,' *International Journal of Urban and Regional Research*, 11, 4, December: 478–99.

Bettelheim, Charles (1974) *Cultural Revolution and Industrial Organization in China*, New York: Monthly Review.

—— (1975) *Economic Calculation and Forms of Property*, New York: Monthly Review.

—— (1976) *Class Struggles in the USSR*, Vol. I, New York: Monthly Review Press.

—— (1978) *Class Struggles in the USSR*, Vol. II, New York: Monthly Review Press.

Black Sash (1989) *Nearly an A-Z Guide to Homelessness on the Witwatersrand*, Black Sash Tvl. Region Urban Removals and Homelessness Group and Community Research and Information Network.

Blumenfeld, Hans (1971) *The Modern Metropolis*, Cambridge, Mass.: MIT Press.

—— (1978) 'Urban planning in the Soviet Union and China,' *Monthly Review*, 30, 2, June: 58–64.

—— (1979) *Metropolis . . . and Beyond*, New York: John Wiley.

Bollens, John C. and Henry J. Schmandt (1970) *The Metropolis: Its People, Politics, and Economic Life*, New York: Harper & Row.

Borja, Jordi (1988) *Estado y Ciudad: Decentralización Politica y Participación*, Barcelona: Promociones y Publicaciones Universitarias.

Bose, Ashish (1980) *India's Urbanization 1901–2001*, New Delhi: Tata McGraw-Hill.

Boyer, Christine (1983) *Dreaming the Rational City*, Cambridge, Mass.: MIT Press.

Boyte, Harry (1979) 'A democratic awakening,' *Social Policy*, 10, 2, September/October: 8–15.

Bratt, Rachel G., Chester Hartman and Ann Meyerson, Eds. (1986) *Critical Perspectives on Housing*, Philadelphia: Temple University Press.

Bromley, Ray and Chris Gerry, Eds. (1979) *Casual Work and Poverty in Third World Cities*, Wiley: Chichester, 3–23.

Brotchie, John F., Peter Hall and Peter W. Newton, Eds. (1987) *The Spatial Impact of Technological Change*, London: Croom Helm.

Burgess, Ernest W. (1961) 'The growth of the city: an introduction to a research project,' in Theodorson: 37–44.

Cadman, David and Geoffrey Payne, Eds. (1990) *The Living City: Towards a Sustainable Future*, London: Routledge.

Caldwell, Malcolm (1977) *The Wealth of Some Nations*, London: Zed Press.

Callow, Alexander B., Jr., Ed. (1969) *American Urban History: An Interpretive Reader with Commentaries*, New York: Oxford University Press.

Caplow, Theodore and Robert Forman (1974) 'Neighborhood interaction in a homogenous community,' in Greer and Greer: 21–36.

Cardoso, Fernando Henrique (1972) 'Dependency and development in Latin America,' *New Left Review*, 74, July–August: 83–95.

Carey, George W. (1981) 'New York: world economy, feudal politics,' *Focus*, 31, 4, March/April: 16pp.

Caro, Robert A. (1974) *The Power Broker: Robert Moses and the Fall of New York*, New York: Knopf.

Castells, Manuel (1977) *The Urban Question: A Marxist Approach*, Cambridge, Mass.: MIT Press.

—— (1983) *The City and the Grassroots*, Berkeley: University of California Press.

—— (1989) *The Informational City*, London: Basil Blackwell.

——, L. Goh and R. Yin-Wang Kwok (1990) *The Shek Kip Mei Syndrome: Economic Development and Public Housing in Hong Kong and Singapore*, London: Pion.

Cecchini, Domenico (1988) 'Le aree urbane in Italia: Scopi, metodi e primi risultati di una ricerca,' *Rivista Economica del Mezzogiorno*, 2, 1: 39–70.

Central Office of Information (1972) *The New Towns of Britain*, London: British Information Services.

Chajcenko, Leonid (1989) 'Sei monologhi,' *Metamorfosi*, 11: 42–52.

Champion, A.G., Ed. (1989) *Counterurbanization: The Changing Nature of Population Deconcentration*, London: Edward Arnold.

Chandler, Tertius and Gerald Fox (1974) *3000 Years of Urban Growth*, New York: Academic Press.

Chase-Dunn, Christopher (1984) 'Urbanization in the world-system: new directions for research,' in Michael Peter Smith, Ed., *Cities in Transformation: Class, Capital, and the State*, Beverly Hills: Sage.

Christaller, Walter (1966) *Central Places in Southern Germany*, New York: Prentice-Hall.

Ciccone, Filippo (1986) 'Alcune riflessioni sull' applicazione della 167,' in Filippo Ciccone, Ed., *Città Pubblica e Qualità Urbana*, Rome: CRESME.

Clark, Kenneth B. (1965) *Dark Ghetto: Dilemmas of Social Power*, New York: Harper & Row.

Clark, Thomas A. (1979) *Blacks in Suburbs: A National Perspective*, New Brunswick: Rutgers University Press.

Clavel, Pierre (1986) *The Progressive City: Planning and Participation 1969–1984*, New Brunswick: Rutgers University Press.

Clawson, Marion (1971) *Suburban Land Conversion in the United States: An Economic and Governmental Process*, Baltimore: Resources for the Future, Inc. and the Johns Hopkins Press.

Clay, Phillip L. and Robert M. Hollister, Eds. (1983) *Neighborhood Policy and Planning*, Lexington, Mass.: Lexington Books.

Cockburn, Cynthia (1982) 'Local government as local state,' in Paris: 175–200.

Cohen, R.B. (1981) 'The new international division of labor, multinational corporations and urban hierarchy,' in Dear and Scott: 287–318.

Collins, George R. and Christiane C. Collins (1965) *Camillo Sitte and the Birth of Modern City Planning*, New York: Random House.

Comune di Roma (1981) *Il PEEP*, Ufficio Speciale del Piano Regolatore, Documenti 3.

—— (1986) *Il Secondo PEEP di Roma*, Ufficio Speciale del Piano Regolatore, Documenti 12.

Coulter, Philip B. (1968) *Politics of Metropolitan Areas: Selected Readings*, New York: Crowell.

Cowley, John, Adah Kaye, Marjorie Mayo and Mike Thompson (1977) *Community or Class Struggle?* London: Stage 1.

Crenson, Matthew (1983) *Neighborhood Politics*, Cambridge, Mass.: Harvard University Press.

Crescenzi, Carlo (1965) 'Il piano di zona de Spinaceto e l'attuazione della 167,' *Urbanistica*, 45, December: 85–94.

Critica Marxista (1989) 'La questione meridionale oggi,' Numero intero. 27, 4, Luglio-Agosto.

Currie, Lauchlin (1976) *Taming the Megalopolis: A Design for Urban Growth*, Oxford: Pergamon Press.

Danielson, Michael N. and Jameson W. Doig (1982) *New York: The Politics of Urban Regional Development* [sic], Berkeley: University of California Press.

Datta, Satya (1990) *Third World Urbanization: Reappraisals and New Perspective*, Stockholm: Swedish Council for Research in the Humanities and Social Sciences.

Davidoff, Paul (1965) 'Advocacy and pluralism in planning,' *Journal of the American Institute of Planners*, 31, 4, November: 331–8.

Davidoff, Paul and Thomas Reiner (1962) 'A choice theory of planning,' *Journal of American Institute of Planners*, 28, May: 331–8.

Dear, Michael and Allen J. Scott, Eds. (1981) *Urbanization and Urban Planning in Capitalist Society*, London: Methuen.

deJanvry, Alain (1981) *The Agrarian Question and Reformism in Latin America*, Baltimore: Johns Hopkins.

Delgado Silva, Angel (1982) *Municipio, Decentralización y Movimiento Popular: Las Alternativas de Izquierda*, Lima. (Unpublished pamphlet.)

de Soto, Hernando (1989) *The Other Path: The Invisible Revolution in the Third World*, New York: Harper & Row.

Dobriner, William Mann (1963) *Class in Suburbia*, Englewood Cliffs, NJ: Prentice-Hall.

Dogan, Mattei and J.D. Kasarda, Eds. (1988) *The Metropolis Era*, Vols. 1–2. Newbury Park: Sage.

Dolbeare, Cushing (1983) 'The low-income housing crisis,' in Chester Hartmann, Ed., *America's Housing Crisis: What is to be Done?* Washington, D.C.: Institute for Policy Studies: 29–75.

—— (1986) 'How the income tax system subsidizes housing for the affluent,' in Bratt, Hartman and Meyerson: 264–71.

Douglass, Mike (1988) 'The transnationalization of urbanization in Japan,' *International Journal of Urban and Regional Research*, 12, 3, September: 343–55.

Downs, Anthony (1970) *Urban Problems and Prospects*, Chicago: Rand McNally.

Dwyer, D.J. (1979) *People and Housing in Third World Cities*, London: Longman.

Edel, Matthew, Elliott D. Sclar and Daniel Luria (1984) *Shaky Places: Homeownership and Social Mobility in Boston's Suburbanization*, New York: Columbia University Press.

Ehrlich, Paul (1968) *The Population Bomb*, New York: Ballantine.

Elliott, Lawrence (1983) *The Little Flower: The Life and Times of Fiorello LaGuardia*, New York: William Morrow & Co.

Engels, Friedrich (1973) *Conditions of the Working Class in England*, Moscow: Progress.

—— (1975) *The Housing Question*, Moscow: Progress.

Epstein, David G. (1973) *Brasilia: Plan and Reality*, Berkeley, University of California Press.

Esser, Josef and Joachim Hirsch (1989) 'The crisis of Fordism and the dimensions of a "post Fordist" regional and urban structure,' *International Journal of Urban and Regional Research*, 13, 3, September: 417–37.

Fainstein, Susan (1987) 'Local mobilization and economic discontent,' in Smith and Feagin: 323–42.

Faludi, Andreas (1973) *A Reader in Planning Theory*, Oxford: Pergamon Press.

Feagin, Joe R. (1986) 'Urban real estate speculation in the United States: implications for social science and urban planning,' in Bratt, Hartman and Meyerson: 99–118.

Finquelievich, Suzana (1990) 'Technology and the urban environment: trends and challenges in Latin American cities,' paper presented to the International Congress 'Sustainable habitat on an urbanized planet?' Berlin, March 19-25: 13pp.

Firey, Walter (1945) 'Sentiment and symbolism and ecological variables,' *American Sociological Review*, 10, April: 140–8.

Fischer, Claude S., Mark Baldassare and Richard J. Ofshe (1975) 'Crowding studies and urban life: a critical review,' *Journal of American Institute of Planners*, 41, 6, November: 406–18.

Fisher, Jack C. (1962) 'Planning the city of socialist man,' *Journal of the American Institute of Planners*, 28, 4, November: 251–65.

——, Ed., (1966) *City and Regional Planning in Poland*, New York: Cornell University Press.

——, Zygmunt Pioro and Milos Savic (1965) 'Socialist city planning: a reexamination,' *Journal of the American Institute of Planners*, 31, 1, February: 31–42.

Fishman, Robert (1987) *Bourgeois Utopias: The Rise and Fall of Suburbia*, New York: Basic Books.

Fitch, Robert (1976) 'Planning New York,' in Roger E. Alcaly and David Mermelstein, Eds., *The Fiscal Crisis of the State*, New York: Vintage, 246–84.

Folin, Marino (1981) 'The production of the general conditions of social production and the role of the state,' in Harloe and Lebas: 51–60.

Forbes, Dean and Nigel Thrift, Eds. (1987) *The Socialist Third World: Urban Development and Territorial Planning*, Oxford: Basil Blackwell.

Forester, John (1989) *Planning in the Face of Power*, Berkeley: University of California Press.

Fortune, Editors of (1958) *The Exploding Metropolis*, Garden City, NY: Doubleday Anchor.

Fox, Geoffrey, Ed. (1989) 'The homeless organize,' special issue of *NACLA*, 23, 4, November/December.

French, R.A. (1984) 'Moscow, the socialist metropolis,' in Sutcliffe: 355–80.
—— and F.E. Ian Hamilton, Eds. (1979) *The Socialist City: Spatial Structure and Urban Policy*, New York: Wiley.
Friedan, Bernard J. and Lynne B. Sagalyn (1989) *Downtown, Inc.: How America Rebuilds Cities*, Cambridge, Mass.: MIT Press.
Freidmann, John (1966) *Regional Development Policy: A Case Study of Venezuela*, Cambridge, Mass.: MIT Press.
—— and C. Weaver (1979) *Territory and Function: The Evolution of Regional Planning*, London: Edward Arnold.
Frobel, F., J. Henrichs and O. Kreye (1980) *The New International Division of Labor*, Cambridge: Cambridge University Press.
Fuerst, J.S., Ed. (1974) *Public Housing in Europe and America*, New York: John Wiley.
Gans, Herbert (1962) *The Urban Villagers*, New York: Free Press.
—— (1968) *People and Plans*, New York: Basic Books: 34–52.
Garrau, Pietro (1990) 'Urbanization and sustainable development.' paper presented to the International Congress 'Sustainable habitat on an urbanized planet?' Berlin, March 19–25: 15pp.
Garreau, Joel (1991) *Edge City: Life on the New Frontier*, New York: Doubleday.
Gartner, Alan, Colin Greer and Frank Riessman (1982) *What Reagan is Doing to Us*, New York: Harper & Row.
Geddes, Patrick (1915) *Cities in Evolution: An Introduction to the Town Planning Movement and to the Study of Civics*, London: Williams and Norgate.
George, Henry (1880) *Progress and Poverty*, New York: Appleton.
Geyer, Hermanus S. (1989) 'Apartheid in South Africa and industrial deconcentration in the PWV area,' *Planning Perspectives*, 4: 251–69.
Gibelli, Maria Cristina (1987) 'Dinamica dello sviluppo urbano in Cina: I problemi attuali e le sfide del futuro,' *Archivio de Studi Urbani e Regionali*, 29: 119–56.
Gilbert, Alan (1989) 'Moving the capital of Argentina: a further example of utopian planning?' *Cities*, 6, 3, August: 234–42.
—— and Josef Gugler (1984) *Cities, Poverty and Development: Urbanization in the Third World*, Oxford: Oxford University Press.
Gilderbloom, John I. and Richard P. Appelbaum (1988) *Rethinking Rental Housing*, Philadelphia: Temple University Press.
Glaab, Charles N. (1963) *The American City: A Documentary History*, Homewood, Ill.: The Dorsey Press.
Glass, Ruth (1989) *Clichés of Urban Doom and Other Essays*, London: Basil Blackwell.
Glazer, Nathan and Daniel P. Moynihan (1970) *Beyond the Melting Pot: The Negroes, Puerto Ricans, Jews, Italians, and Irish of New York City*, Cambridge, Mass.: MIT Press.
Gonzalez Casanova, Pablo (1970) *Sociologia de la Explotación*, 2nd Ed., Mexico City: Siglo XXI Editores.
Goodman, Paul and Percival Goodman (1960) *Communitas: Means of Livelihood and Ways of Life*, New York: Vintage.
Gordon, David (1978) 'Capitalist development and the history of American

cities,' in Tabb and Sawers: 25–63.
Gorz, Andre (1985) *Paths to Paradise: On the Liberation from Work*, Boston: South End Press.
Gottdeiner, M. (1985) *The Social Production of Urban Space*, Austin: University of Texas Press.
—— (1987) *The Decline of Urban Politics: Political Theory and the Crisis of the Local State*, Newbury Park: Sage.
Gottmann, Jean (1961) *Megalopolis: The Urbanized Northeastern Seaboard of the United States*, Cambridge, Mass.: The Twentieth Century Fund.
—— (1989) 'Past and present of the western city,' in Lawton: 168–76.
Gramsci, Antonio (1957) *La Questione Meridionale*, Rome: Riuniti.
Gratz, Roberta Brandes (1989) *The Living City*, New York: Simon & Schuster.
Greer, Scott (1962) *Governing the Metropolis*, New York: Wiley.
—— and Ann Lennarson Greer, Eds. (1974) *The Neighborhood and Ghetto: The Local Area in Large-Scale Society*, New York: Basic Books.
Gugler, Josef, Ed. (1988) *The Urbanization of the Third World*, Oxford: Oxford University Press.
Gurley, John G. (1976) *China's Economy and the Maoist Strategy*, New York: Monthly Review Press.
Gutnov, Alexei, *et al.* (1968) *The Ideal Communist City*, New York: Braziller.
Habitat (1987) *Global Report on Human Settlements*, London: Oxford University Press.
Hadjimichalis, Costis (1987) *Uneven Development and Regionalism: State Territory and Class in Southern Europe*, Beckenham: Croom Helm.
Hall, P. and P. Preston (1988) *The Carrier Wave: New Information Technology and the Geography of Innovation 1846–2003*, London: Unwin Hyman.
Hall, Peter (1966) *The World Cities*, New York: McGraw-Hill.
—— (1982) *Urban and Regional Planning*, London: Allen & Unwin.
—— (1984) 'Metropolis 1890–1940: challenges and responses,' in Sutcliffe: 19–66.
—— (1989) 'The rise and fall of great cities: economic forces and population responses,' in Lawton: 20–31.
—— and Dennis Hay (1980) *Growth Centres in the European Urban System*, London: Heinemann.
Hamnett, Chris and Bill Randoph (1988) *Cities, Housing and Profits: Flat Break-up and the Decline of Private Renting*, London: Hutchinson.
Hanley, Robert (1991) 'Ethnic mosaic grows beyond New York', *New York Times*, April 17: B1–2.
Hardoy, Jorge (1982) 'The building of Latin American cities,' in Alan Gilbert, Ed., *Urbanization in Contemporary Latin America*, New York: Wiley, 19–34.
—— and David Satterthwaite (1986) 'Urban change in the Third World: are recent trends a useful pointer to the urban future?' *Habitat International*, 10, 3: 33–52.
Harloe, Michael and Elizabeth Lebas, Eds. (1981) *City, Class and Capital*, London: Edward & Arnold.
Harris, Nigel (1986) *The End of the Third World: Newly Industrializing*

Countries and the Decline of an Ideology, London: Penguin.

Hartman, Chester, Dennis Keating and Richard LeGates (1982) *Displacement: How to Fight It*, Berkeley: National Housing Law Project.

Harvey, David (1973) *Social Justice and the City*, Baltimore: Johns Hopkins University Press.

—— (1989) *The Urban Experience*, Baltimore: Johns Hopkins University Press.

Hauser, Philip M. and Leo F. Schnore, Eds. (1965) *The Study of Urbanization*, New York: Wiley.

Hayden, Dolores (1984) *Redesigning the American Dream*, New York: W.W. Norton.

Henderson, J. and M. Castells, Eds. (1987) *Global Restructuring and Territorial Development*, London: Sage.

Herbers, John (1987) 'Black poverty spreads in 50 biggest U.S. cities,' *New York Times*, January 26: A27.

Hester, Randoph T. (1984) *Planning Neighborhood Space with People*, New York: Van Nostrand.

Higgins, Bryan R. (1990) 'Geographical revolutions and revolutionary geographies: nature, space and place in the urban development of Nicaragua,' paper presented to the International Congress 'Sustainable habitat on an urbanized planet?' Berlin, March 19–25: 21pp.

Homefront (1977) *Housing Abandonment in New York City*, New York.

Hopper, Kim and Jill Hamberg (1986) 'The making of America's homeless: from skid row to new poor, 1945–1984', in Bratt, Hartman and Meyerson, Eds.: 12–40.

Hoover, Edgar M. (1968) 'The evolving form and organization of the metropolis,' in Harvey S. Perloff and Lowdon Wingo, Jr., Eds., *Issues in Urban Economics*, Baltimore: Johns Hopkins Press: 237–84.

Horan, James F. and G. Thomas Taylor, Jr. (1977) *Experiments in Metropolitan Government*, New York: Praeger.

Hoselitz, Bert F. (1969) 'The role of cities in the economic growth of underdeveloped countries,' in Gerald Breese, Ed., *The City in Newly Developing Countries*, Princeton: Princeton University Press: 232–45.

Howard, Ebenezer (1965) *Garden Cities of Tomorrow*, Ed. by F.J. Osborn. London: Faber & Faber.

Hughes, Thomas P. and Agatha C. Hughes (1990) *Lewis Mumford: Public Intellectual*, New York: Oxford University Press.

Hurd, Richard M. (1903) *Principles of City Land Values*, New York: The Record and Guide.

Illeris, Sven (1987) 'The metropolis, the periphery and the Third Wave,' *Town Planning Review*, 58, 1, January: 19–28.

Institut d'Estudis Metropolitans de Barcelona (1988) *Cities: Statistical, Administrative and Graphical Information on the Major Urban Areas of the World*, Vols. I-IV, Barcelona.

Isaacs, Reginald (1948) 'The "neighborhood unit" is an instrument of segregation,' *Journal of Housing*, 5: 215–19.

Jackson, Kenneth (1985) *Crabgrass Frontiers: The Suburbanization of the United States*, New York: Oxford University Press.

Jacobs, Jane (1961) *The Death and Life of Great American Cities*, New York:

Vintage.
—— (1984) *Cities and the Wealth of Nations: Principles of Economic Life*, New York: Vintage.
Johnson, James H. (1972) *Urban Geography: An Introductory Analysis*, Oxford: Pergamon.
Jones, Delmos J. (1979) 'Not in my community: the neighborhood movement and institutionalized racism,' *Social Policy*, 10, 2, September/October: 44–6.
Jones, Emrys (1990) *Metropolis: The World's Great Cities*, New York: Oxford University Press.
Kain, John F. (1973) 'Effect of housing market segregation on urban development,' in Jon Pynoos, Robert Schafer and Chester W. Hartman, Eds., *Housing Urban America*, Chicago: Aldine: 251–66.
Kanter, Rosabeth Moss (1972) *Commitment and Community: Communes and Utopias in Sociological Perspective*, Cambridge, Mass.: Harvard University Press.
Katznelson, Ira (1981) *City Trenches: Urban Politics and the Patterning of Class in the United States*, New York: Pantheon.
Keating, Dennis, Keith P. Rasey and Norman Krumholz (1990) 'Community development corporations in the United States: their role in housing and urban development,' in Willem van Vliet and Jan van Weesep, Eds., *Government and Housing Developments in Seven Countries*, Sage Publications Urban Affairs Annual Reviews, 36: 206–19.
Keating, Michael (1988) *State and Regional Nationalism: Territorial Politics and the European State*, New York: Harvester.
Keller, Suzanne (1968) *The Urban Neighborhood: A Sociological Perspective*, New York: Random House.
Kelley, Klara Bonsack (1976) 'Dendritic central place systems and the regional organization of Navajo trading posts,' in C.A. Smith: 219–54.
Kerblay, M.B. (1968) 'Moscou,' *Notes et Etudes Documentaires*, 3, 493, May 24: 79pp.
King, Anthony D. (1990) *Urbanism, Colonialism and the World Economy: Cultural and Spatial Foundations of the World Urban System*, London: Routledge.
Kleinman, David (1981) *Human Adaptation and Population Growth: A Non-Malthusian Perspective*, Totowa, NJ: Allenheld, Osmun.
Kling, Joseph M. and Prudence S. Posner, Eds. (1990) *Dilemmas of Activism: Class, Community and the Politics of Local Mobilization*, Philadelphia: Temple University Press.
Kotler, Milton (1969) *Neighborhood Government: The Local Foundations of Political Life*, Indianapolis: Bobbs-Merrill.
Kozhurin, V.S. and S.A. Pogodin (1981) 'Changes in the urban population of the USSR, 1939–1979,' *The Soviet Review*: 3–12.
Kuklinski, Antoni, Ed. (1972) *Growth Poles and Growth Centres in Regional Planning*, London: Mouton.
Kwok, R. Yin-Wang (1987) 'Recent urban policy and development in China: a reversal of "anti-urbanism",' *Town Planning Review*, 58, 4, October: 383–99.
Laska, Shirley Bradway and Daphne Spain, Eds. (1980) *Back to the City: Issues in Neighborhood Renovation*, New York: Pergamon Press.

Latoar, Alessandra (1989) 'Note sui nuovi indirizzi progettuali a Mosca,' *Metamorfosi*, 11: 8–9.

Lawson, Ronald (1986) *The Tenant Movement in New York City*, New Brunswick: Rutgers University Press.

Lawton, Richard, Ed. (1989) *The Rise and Fall of Great Cities: Aspects of Urbanization in the Western World*, London: Belhaven.

Leacock, Eleanor Burke, Ed. (1971) *The Culture of Poverty: A Critique*, New York: Simon & Schuster.

Leavitt, Helen (1970) *Superhighway - Superhoax*, Garden City, NY: Doubleday.

LeGates, Richard T. and Chester Hartman (1981) 'Displacement', *Clearinghouse Review*, 15, 3, July: 207–49.

Lindblom, Charles (1959) 'The science of muddling through,' 30, Spring: 79–88.

Lipton, Michael (1977) *Why Poor People Stay Poor: Urban Bias in World Development*, Cambridge, Mass.: Harvard University Press.

Logan, John R. and Harvey L. Molotch (1987) *Urban Fortunes: The Political Economy of Place*, Berkeley: University of California Press.

Long, Larry (1988) *Migration and Residential Mobility in the United States*, New York: Russell Sage Foundation.

Lowder, Stella (1986) *Inside Third World Cities*, Beckenham: Croom Helm.

Lowe, Marcia D. (1990) 'Alternatives to the automobile: transport for livable cities,' Worldwatch Paper 98, October: 49pp.

Lozano, Eduardo E. (1990) *Community Design and the Culture of Cities: The Crossroad and the Wall*, Cambridge: Cambridge University Press.

Lynch, Kevin (1960) *The Image of the City*, Cambridge, Mass.: MIT Press.

McGuire, Chester C. (1981) *International Housing Policies*, Lexington, Mass.: Lexington Books.

McKelvey, Blake (1968) *The Emergence of Metropolitan America 1915–1966*, New Brunswick: Rutgers University Press.

Magnaghi, Alberto (1982) *Il Sistema di Governo delle Regioni Metropolitane*, Milan: Franco Angeli.

Mannheim, Karl (1940) *Man and Society in an Age of Reconstruction*, New York: Harcourt, Brace & World.

Marcelloni, Maurizio, Ed. (1987) *Il Regime dei Suoli in Europa*, Milan: Franco Angeli.

Marcuse, Peter (1986) 'Abandonment, gentrification and displacement: the linkages in New York City,' in Smith and Williams: 153–77.

Marcuse, Peter (1988) 'A man for all systems,' *Monthly Review*, 40, 6, November: 43–8.

Mariátegui, Jose Carlos (1928) *Siete Ensayos de Interpretación de la Realidad Peruana*, Lima: Amauta.

Markusen, Ann, Peter Hall and Amy Glasmeier (1986) *High Tech America*, Boston: Allen & Unwin.

Marquit, Erwin (1983) *The Socialist Countries: General Features of Political, Economic and Cultural Life*, Minneapolis: MEP Publications.

Marris, Peter (1960) 'Slum clearance and family life in Lagos,' *Human Organization*, 19: 123–8.

Marx, Karl (1967) *Capital*, Vol. I, New York: International Publishers.

—— (1973) *Grundrisse*, New York: Vintage.

—— and Friedrich Engels (1968) *Selected Works*, New York: International Publishers.

Masotti, Louis H. and Jeffrey K. Hadden (1973) *The Urbanization of the Suburbs*, Beverly Hills: Sage.

Mathey, Kosta (1988) 'A Cuban interpretation of self-help housing,' *Trialog*, 18, 3, Quartal: 24–30.

Mazziotti, Donald F. (1982) 'The underlying assumptions of advocacy planning: pluralism and reform,' in Paris: 207–26.

Meadows, Donella, *et al.* (1974) *The Limits to Growth*, New York: Signet.

Medvedkov, Olga (1990) *Soviet Urbanization*, London: Routledge.

Meyer, J.R., J.F. Kain and M. Wohl (1965) *The Urban Transportation Problem*, Cambridge, Mass.: Harvard University Press.

Meyers, William (1985) 'The nonprofits drop the "non" ', *New York Times*, November 24.

Meyerson, Martin (1956) 'Building the middle range bridge for comprehensive planning,' *Journal of the American Institute of Planners*, 22, 2, Spring: 58–64.

—— and Edward C. Banfield (1955) *Politics, Planning and the Public Interest*, New York: Macmillan.

Miliutin, N.A. (1974) *Sotsgorod*, Cambridge, Mass.: MIT Press.

Mills, C. Wright (1959) *The Sociological Imagination*, London: Oxford University Press.

Miner, Horace (1952) 'The folk-urban continuum,' *American Sociological Review*, 17, October: 529–37.

Mingione, Enzo (1983) *Urbanizzazione, Classi Sociali, Lavoro Informale*, Milan: Franco Angeli.

Mollenkopf, John (1983) *The Contested City*, Princeton: Princeton University Press.

Morris, David and Karl Hess (1965) *Neighborhood Power*, Boston: Beacon Press.

Morris, Pauline (1981) *A History of Black Housing in South Africa*, Johannesburg: South Africa Foundation.

Mumford, Lewis (1961) *The City in History*, New York: Harcourt, Brace & World.

Musil, Jiri (1980) *Urbanization in Socialist Countries*, White Plains, NY: M.E. Sharpe.

Nash, June and Helen Safa, Eds. (1976) *Sex and Class in Latin America*, New York: Praeger.

National Urban Coalition (1978) *Displacement: City Neighborhoods in Transition*, Washington: National Urban Coalition.

—— (1979) Neighborhood Transition without Displacement, Washington, D.C.: National Urban Coalition.

Newsweek (1990) 'After apartheid,' April 2: 34–5.

Nkrumah, Kwame (1965) *Neo-Colonialism: The Last Stage of Imperialism*, New York: International Publishers.

Odmann, Ella and Gun-Britte Dahlberg (1970) *Urbanization in Sweden*, Stockholm: Allmanna Forlaget.

Orfield, Gary (1986) 'Minorities and suburbanization,' in Bratt, Hartman and Meyerson: 221–9.

Osborn, F.J. (1941) 'Reflections on Density,' *Town and Country Planning*, 9, 35, Autumn: 121–6, 146.

Osborn, Sir Frederic and Arnold Whittick (1969) *The New Towns: The Answer to Megalopolis*, London: Leonard Hall.

Osborne, Robert J. (1966) 'How the Russians plan their cities,' *Trans-Action*, 3, 6, September–October: 25–30.

Osofsky, Gilbert (1971) *Harlem: The Making of a Ghetto*, New York: Harper & Row.

Palloix, Christian (1977) 'The self-expansion of capital on a world scale,' *Review of Radical Political Economy*, 9, 2: 1–28.

Panos Institute (1990) *We Speak for Ourselves: Social Justice, Race and Environment*, Washington, D.C.

Paris, Chris, Ed. (1982) *Critical Readings in Planning Theory*, Oxford: Pergamon.

Park, Robert E., Ernest W. Burgess and R.D. McKenzie (1925) *The City*, Chicago: University of Chicago Press.

Parker, John A., *et al.* (1952) *An Examination of Soviet Theory and Practice in City and Regional Planning*, Chapel Hill: Institute for Research in Social Science, University of North Carolina.

Perlman, Janice (1976) *The Myth of Marginality*, Berkeley: University of California Press.

Perloff, Harvey (1980) *Planning the Post-Industrial City*, Washington, D.C.: APA Planners Press.

Perry, Clarence (1929) *The Neighborhood Unit*, Regional Survey of New York and its Environs, Vol. 7.

Perulli, Paulo, Ed. (1989) *Città della Scienza e della Technologia*, Quaderni della Fondazione Istituto Gramsci Veneto, 6/7: Venezia.

Pinch, Steven (1985) *Cities and Services: The Geography of Collective Consumption*, London: Routledge & Kegan Paul.

Piven, Francis Fox and Richard A. Cloward (1971) *Regulating the Poor: The Functions of Social Welfare*, New York: Random House.

—— (1979) *Poor People's Movements: Why they Succeed, How they Fail*, New York: Pantheon.

Plotkin, Sidney (1987) *Keep Out: The Struggle for Land Use Control*, Berkeley: University of California Press.

Portes, Alejandro (1985) 'Latin American class structures: their composition and change during the last decades,' *Latin American Research Review*, 20, 3: 7–40.

——, Manuel Castells and Lauren A. Benton, Eds. (1989) *The Informal Economy: Studies in Advanced and Less Developed Countries*, Baltimore: Johns Hopkins University Press.

Preston, Samuel H. (1988) 'Urban growth in developing countries: a demographic reappraisal,' in Gugler: 11–32.

Redfield, Robert (1947) 'The folk society,' *The American Journal of Sociology*, 52, January: 293–308.

Reps, John W. (1964) 'Requiem for zoning,' in H. Wentworth Eldredge, Ed., *Taming Megalopolis*, Garden City, NY: Anchor: 746–59.

—— (1965) *The Making of Urban America: A History of City Planning*, Princeton: Princeton University Press.

Ridgeway, James (1970) *The Politics of Ecology*, New York: E.P. Dutton.

Rodwin, Lloyd (1969) *Planning Urban Growth and Regional Development: The Experience of the Guayana Program in Venezuela*, Cambridge, Mass.: MIT Press.

Rohe, William M. amd Lauren B. Gates (1985) *Planning with Neighborhoods*, Chapel Hill: University of North Carolina Press.

Romero, Emilio and Cesar Levano (1969) *Regionalismo y Centralismo*, Lima: Amauta.

Rondinelli, Dennis (1983) *Secondary Cities in Developing Countries: Policies for Diffusing Urbanization*, Beverly Hills: Sage.

—— and Kenneth Ruddle (1978) *Urbanization and Rural Development: A Spatial Policy for Equitable Growth*, New York: Praeger.

Rudofsky, Bernard (1969) *Streets for People: A Primer for Americans*, Garden City, NY: Anchor Press/Doubleday.

Rudolph, Joseph R. and Robert J. Thompson, Eds. (1989) *Ethnoterritorial Politics, Policy, and the Western World*, Boulder: Lynne Reiner.

Rybczynski, Witold (1991) ' "Edge Cities": the people's answer to planners,' *New York Times*, November 17: 36.

Sale, Kirkpatrick (1985) *Dwellers in the Land: The Bioregional Vision*, San Francisco: Sierra Club Books.

Sassen, Saskia (1988) *The Mobility of Labor and Capital: A Study in International Investment and Labor Flow*, Cambridge: Cambridge University Press.

—— (1991) *The Global City: New York, London, Tokyo*, Princeton: Princeton University Press.

Saunders, Peter (1986) *Social Theory and the Urban Question*, New York: Holmes & Meier.

Sawers, Larry (1977) 'Urban planning in the Soviet Union and China,' *Monthly Review*, 28, 10, March: 34–48.

Schaeffer, K.H. and Elliot Sclar (1975) *Access for All: Transportation and Urban Growth*, Harmondsworth: Penguin.

Schoenberger, Ed. (1988) 'From Fordism to flexible accumulation: technology, competitive strategies and international location,' *Environment and Planning D Society and Space*, 6: 245–62.

Schumacher, E.F. (1973) *Small is Beautiful*, New York: Harper.

Schuman, Tony (1991) 'A framework for existence: toward a non-heroic architecture,' *Journal of Architectural and Planning Research*, 8: 4–8.

Schwartz, David C., Richard C. Ferlauto and Daniel N. Hoffman (1988) *A New Housing Policy for America: Recapturing the American Dream*, Philadelphia: Temple University Press.

Scott, A.J. (1988) 'Flexible production systems and regional development: the rise of new industrial spaces in North America and Western Europe,' *International Journal of Urban and Regional Research*, 12, 2, June: 171–86.

Scott, Allen J. (1980) *The Urban Land Nexus and the State*, London: Pion.

Scott, Mel (1969) *American City Planning since 1890*, Berkeley: University of California Press.

Segre, Roberto (1988) *Arquitectura y Urbanismo Modernos*, La Habana: Editorial Arte y Literatura.

Self, Peter (1982) *Planning the Urban Region: A Comparative Study of Policies and Organizations*, London: Allen & Unwin.

Sharpe, L.J., Ed. (1979) *Decentralist Trends in Western Democracies*, London: Sage.

Short, John R. (1989) *The Humane City*, London: Basil Blackwell.

Sistema Permanente di Servizi (1991) *Le Città Metropolitane*, Rome: SPS.

Sitte, Camillo (1965) *City Planning According to Artistic Principles*, New York: Random House.

Sjoberg, Gideon (1960) *The Preindustrial City*, New York: Free Press.

Smerk, George M., Ed. (1968) *Readings in Urban Transportation*, Bloomington: Indiana University Press.

Smith, B.C. (1985) *Decentralization: The Territorial Dimension of the State*, Boston: Allen & Unwin.

Smith, Carol A., Ed. (1976) *Regional Analysis*, New York: Academic Press.

Smith, Dennis (1988) *The Chicago School: A Liberal Critique of Capitalism*, London: Macmillan.

Smith, Michael Peter (1988) *City State and Market: The Political Economy of Urban Society*, New York: Basil Blackwell.

—— and Joe R. Feagin, Eds. (1987) *The Capitalist City: Global Restructuring and Community Politics*, Oxford: Basil Blackwell.

—— and Richard Tardanico (1987) 'Urban theory reconsidered: production, reproduction and collective action,' in Smith and Feagin: 87–110.

Smith, Neil (1984) *Uneven Development: Nature, Capital and the Production of Space*, Oxford: Basil Blackwell.

—— (1986) 'On the necessity of uneven development,' *International Journal of Urban and Regional Research*, 10, 1, March: 87–104.

—— and Peter Williams, Eds. (1986) *Gentrification of the City*, Boston: Allen & Unwin.

Snell, Bradford C. (1973) *American Ground Transport: A Proposal for Restructuring the Automobile, Truck, Bus and Rail Industries*, Washington, D.C.: Government Printing Office.

Sofen, Edward (1966) *The Miami Metropolitan Experience*, Garden City, NY: Anchor Books.

Sorel, Georges (1969) *The Illusions of Progress*, Berkeley: University of California Press.

Squires, Gregory D., *et al.* (1989) *Chicago: Race, Class and the Response to Urban Decline*, Philadelphia: Temple University Press.

Starr, Roger (1976) 'City's housing administrator proposes planned shrinkage of some slums,' *New York Times*, February 3.

Stave, Bruce M., Ed. (1975) *Socialism and the Cities*, Port Washington, NY: Kennikat Press.

Stein, Jay M., Ed. (1988) *Public Infrastructure Planning and Management*, Newbury Park: Sage.

Stepanov, Vyacheslav K. (1974) 'Union of Soviet Socialist Republics,' in Arnold Whittick, Ed., *Encyclopedia of Urban Planning*, New York: McGraw-Hill: 1135–58.

Sternlieb, George (1966) *The Tenement Landlord*, New Brunswick: Rutgers University Press.

Stevens, William K. (1991) 'At meeting on global warming, U.S. stands alone,' *New York Times*, September 10: C1, 9.

Stockholm City Building Department (1972) *Stockholm: Urban Environment*, Stockholm.

Stone, Michael E. (1986) 'Housing and the dynamics of U.S. capitalism,' in Bratt, Hartman and Meyerson: 41-67.

Studies in Comparative Communism (1979) 'The urban challenge: cities under communism,' special issue, 12, 4, Winter.

Sundaram, K.V. (1977) *Urban and Regional Planning in India*, New Delhi: Vikas Publishing House.

Susman, Paul (1987) 'Spatial equality and socialist transformation in Cuba,' in Forbes and Thrift: 250–81.

Sutcliffe, Anthony, Ed. (1984) *Metropolis: 1890-1940*, London: Mansell.

Sutcliffe, Michael, Alison Todes and Norah Walker (1990) 'Managing the cities: an examination of state urban policies since 1986,' paper presented to the International Conference 'Sustainable habitat on an urbanized planet?' Berlin, March 19–25: 20pp.

Szelenyi, Ivan, Ed. (1984) *Cities in Recession: Critical Responses to the Urban Policies of the New Right*, London: Sage.

Tabb, William K. and Larry Sawers, Eds. (1978) *Marxism and the Metropolis*, New York: Oxford University Press.

Tanzer, Michael (1974) *The Energy Crisis: World Struggle for Power and Wealth*, New York: Monthly Review Press.

Taub, Richard, D. Garth Taylor and Jan D. Dunham (1987) *Paths of Neighborhood Change: Race and Crime in Urban America*, Chicago: University of Chicago Press.

Taylor, Michael and Nigel Thrift, Eds. (1986) *Multinationals and the Restructuring of the World Economy*, Beckenham: Croom Helm.

Teaford, Jon C. (1986) *The Twentieth Century American City: Problem, Promise, and Reality*, Baltimore: Johns Hopkins Press.

Tennenbaum, Robert (1990) 'Hail, Columbia,' *Planning*, 56, 5, May: 16–17.

Theodorson, George A., Ed. (1961) *Studies in Human Ecology*, New York: Harper & Row.

Thomas, Michael J. (1982) 'The procedural planning theory of A. Faludi,' in Paris: 13–26.

Timberlake, Michael, Ed. (1985) *Urbanization in the World Economy*, New York: Academic Press.

Troen, Ilan (1988) 'The transformation of Zionist planning policy: from rural settlements to an urban network,' *Planning Perspectives*, 3: 3–23.

Turner, James (1970) 'Blacks in the cities: land and self-determination,' *The Black Scholar*, 1, 6, April: 9–13.

Turner, John F.C. and Robert Fichter, Eds. (1972) *Freedom to Build*, New York: Macmillan.

United Nations (1989) *United Nations Demographic Yearbook*, New York.

—— (1989a) *United Nations Energy Statistics Yearbook*, New York.

United States Congress (1977) *How Cities Can Grow Old Gracefully*,

Washington, D.C.: House Committee on Banking, Finance and Urban Affairs.

United States Department of Labor (1971) *Learning from British New Towns*, Washington: Department of Labor.

Urban Planning Aid (1973) *Community Housing Development Corporations: The Empty Promise*, Boston: Urban Planning Aid.

Veblen, Thorsten (1961) *The Theory of the Leisure Class*, New York: Modern Library.

Velázquez, Daniel Rodríguez (1989) 'From neighborhood to nation', in Fox: 22–8.

Velentey, D.I. (1977) *An Outline Theory of Population*, Moscow: Progress Publishers.

Venys, Ladislav and Bohuslav Kohout (1975) Interview at the offices of Terplan, Prague: June.

Violich, Francis, with Robert Daughters (1987) *Urban Planning for Latin America: The Challenge of Metropolitan Growth*, Boston: Gunn & Hain.

Walsh, Annmarie Hauck (1969) *The Urban Challenge to Government: An International Comparison of Thirteen Cities*, New York: Praeger.

Walton, John, Ed. (1985) *Capital and Labour in the Urbanized World*, London: Sage.

—— (1987) 'Urban protest and the global political economy: the IMF riots,' in Smith and Feagin: 364–86.

Warner, Sam Bass (1962) *Street Car Suburbs*, Cambridge, Mass.: Harvard University Press.

Wallerstein, Emmanuel (1974) *The Modern World System*, New York: Academic Press.

Weinstein, James (1968) *The Corporate Ideal in the Liberal State, 1900–1918*, Boston: Beacon Press.

Weisberg, Barry (1971) *Beyond Repair: The Ecology of Capitalism*, Boston: Beacon Press.

Werth, Joel T. and David Bryant (1979) *A Guide to Neighborhood Planning*, Chicago: American Planning Association.

Whyte, William H. (1988) *City: Rediscovering the Center*, New York: Doubleday.

Wilson, James Q. (1966) *Urban Renewal: The Record and the Controversy*, Cambridge, Mass.: MIT Press.

Winner, Langdon (1977) *Autonomous Technology: Technics-out-of-control as a Theme in Political Thought*, Cambridge, Mass.: MIT Press.

Wirth, Louis (1964) *On Cities and Social Life*, Chicago: University of Chicago Press.

Wolf, Eleanor (1963) 'The tipping point in racially changing neighborhoods,' *Journal of the American Institute of Planners*, 29, August: 217–22.

Wolfe, Tom (1981) *From Bauhaus to Our House*, New York: Farrar, Straus & Giroux.

World Bank (1974) *Sites and Services Projects*, Washington, D.C.

World Marxist Review Working Group (1985) 'A "housing crisis" under socialism? Look at the facts', Working Group on Scientific Information and Documentation, *World Marxist Review*, 28, 7, July: 120–4.

Worthy, William (1976) *The Rape of Our Neighborhoods*, New York: William Morrow.
Wu, Chung-Tong (1987) 'Chinese socialism and uneven development,' in Forbes and Thrift: 53–97.
Yergin, Daniel (1991) *The Prize: The Epic Quest for Oil, Money and Power*, New York: Simon & Schuster.
Young, Ken and Patricia L. Garside (1982) *Metropolitan London: Politics and Urban Change 1837-1981*, London: Edward Arnold.
Zeitlin, Morris (1972) *Guide to the Literature of Cities: Abstracts and Bibliography*, Monticello, Ill.: Council of Planning Librarians.
—— (1985) 'Soviet urbanization and urban planning,' *Political Affairs*, 64, 4, April: 27–34.
—— (1990) *American Cities: A Working Class View*, New York: International Publishers.

INDEX